gene in biology are reflected today in the similar primacy of nucleic acids as the basis of life. For students of the history of ideas, a collection of these essays would illustrate how genetic thinking prepared the world view for molecular biologists.

The relation of science to values is often neglected because of the inaccessibility of the written contributions of famous scientists. To read Muller's major essays in these two areas is an important way to evaluate a scientist's career, his maturation of ideas, and his developing application of science to society.

Elof Axel Carlson is a geneticist who studied under Muller at Indiana University where he received his Ph.D. in 1958. Dr. Carlson is professor of biology, State University of New York at Stony Brook. He received the E. Harris Harbison Award for Distinguished Teaching, 1972, from the Danforth Foundation.

Man's Future Birthright

Man's Future Birthright

Essays on Science and Humanity

by H. J. Muller

Edited by Elof Axel Carlson

State University of New York Press Albany, 1973

Elof Axel Carlson is
professor of biology at
the State University of
New York, Stony Brook.

Man's Future Birthright:
Essays on Science and Humanity
by H. J. Muller
First Edition
Published by State University of
New York Press
99 Washington Avenue,
Albany, New York 12210

Printed in the United States of America

Library of Congress Cataloging
in Publication Data

Muller, Hermann Joseph, 1890-1967.
Man's future birthright.

Contents: Glass, B. Introduction.—Possible
advances of the next hundred years: a biologists view.
—Science fiction as an escape.—[Etc.]
1. Human genetics—Social aspects. 2. Biology—
Social aspects. 3. Muller, Hermann Joseph, 1890-1967.
I. Title.
QH431.M942 301.24'3 79-171215
ISBN 0-87395-097-6
ISBN 0-87395-197-2 (microfiche)

Contents

Preface

HERMANN Joseph Muller lived the life of a scientist and of a social reformer. Although he acknowledged that "the winning of the facts" was in itself its own reward, he felt even more rewarded when those facts could be applied to human betterment. As a scientist Muller had a dominant influence on the development of the gene concept and the study of mutation. His theoretical contributions to genetics and evolution provided, in large measure, the world-view which molecular biology has adopted. As a social reformer he sought intellectual freedom and voluntary eugenics, achieving only partial success in his worldwide efforts to popularize these views. He was also a leading spokesman for protecting the populace from careless exposure to radiation.

The essays in this volume reflect Muller's optimism that through education, especially emphasizing the scientific world-view, man would achieve lasting peace, limit his population, achieve a social equity in the wealth of the world, and find adventure in the use of his knowledge to better himself and his progeny. Muller realized that success in achieving these ideals is not an inevitable attribute of human evolution. Rather, man had to strive continuously to reject the ignorance, myths, vested interests, superstitions, and resistance of the past.

Muller's optimism might be challenged today by the wave of pessimism and repudiation of rationalism that emerged as a reaction to the unsolved contemporary problems of militarism, racism, poverty, and pollution. Yet the necessary rational activities to combat these issues has seldom been tried, and the failure of violence, withdrawal, and counter-cultural movements to solve them underscores the need to reexamine rationalist viewpoints, such as those which Muller advocates in this volume.

The topics of these essays reflect Muller's broad range of interests and his insatiable curiosity. He did not shy from the philosopher's task to raise broader questions about the meaning of life and of freedom, nor did he hesitate to explore the future as a legitimate extension of science. He surprised many students of the humanities by applying his genetic analysis to human values and by asserting that these values had a biological basis through natural selection. His insights into the space program, eight years before the moon landing, reveal how thoroughly Muller explored the implications of scientific facts.

After his retirement, Muller had hoped to write a book-length popularization of his philosophic views and his eugenic program. Illness delayed this and many other projects. He died with regrets that these tasks could not be completed by him, but he hoped that a new generation would extend, in ever more creative ways, the guidance of human evolution and the world view of moderns.

A companion volume of Muller's major theoretical essays in science, *The Modern Concept of Nature,* was published by the State University of New York Press.

Elof Axel Carlson

Introduction

FROM his earliest years, as he has personally testified, Hermann Joseph Muller was profoundly concerned about human evolution. His awakening to the idea reached back, he recollected, to his father's explanations of the evolution of life and of man on visits with his young son to the American Museum of Natural History in Manhattan. Later, the concept was extended into the future as the boy revelled in H. G. Wells's science fiction. Any sufficient explanation of the present state by reliance upon a scientific knowledge of the past and of the forces of change and evolution would surely bring about a future in which man, with unimaginable power under his control and vast knowledge tempered by wisdom, would shape the world yet to be. A man's life, above all else, thought the lad, ought to be concentrated on the progressive realization of that dream. That was why, so early, Muller dedicated himself to science, which alone could more fully explain the past and unlock the doors of the future. Of the natural sciences, biology in particular seemed to be most closely and naturally tied to the illumination of the history of man and the future prospect. And so biology claimed the eager youth.

Muller's student years in the fly laboratory at Columbia University were to discipline his vision. Yet more than any others of the famous Morgan group, Muller, I think, was the visionary and the humanist. Those aspects of his personality were never effaced, but on the contrary became strengthened with the years, to form indeed the warp on which his philosophy of life was woven. A deepening concern for such matters permeated his thinking and made his science always a minister to his vision and his ideals. Examine the subjects, if you will, which are treated in this volume and which are a small selection of what might in fact have been assembled from his writings. Here we find a blend of speculation with warnings of personal and social danger, a philosophy of human freedom, a search for values embedded in the evolutionary history of mankind, and a concern for social improvement and controlled evolution toward a higher estate — all of which bear witness to his optimistic belief in a better society and to a biologically improved nature and capacity of man.

My own earliest acquaintance with Hermann Joseph Muller antedated by some months the beginning of my graduate studies at the University of Texas under his direction. The occasion was a meeting of the McLennan County Medical Society in Waco, Texas, where I was completing my master's degree at Baylor University. To an audience of physicians who came expecting to be told only about the exciting discovery made by Muller in the preceding months, that by subjecting them to X rays the genes of the fruit fly *Drosophila* can be made to mutate at rates a thousand times as great as the spontaneous frequency, Muller administered a vigorous shock. After a preliminary account of the discovery, he launched into a discussion of its implications. First he emphasized the frightening dangers which the discovery indicated were being inflicted on the population by physicians' careless employment of X rays in diagnosis and treatment. Thereafter, he elaborated on the possibilities for evolution inherent in an understanding of the mutation process. He left an audience some of whom were stunned, some left in apoplectic anger. It was not to be the last time that Muller's outspoken humanism provoked a similar reaction.

During my years of graduate study under Muller at the University of Texas, his dissatisfaction with the American social and economic system, then oscillating so wildly between boom and inflation, then depression, unemployment, and despair, reached a peak. Muller was led to embrace the Marxian ideal of an ordered society and to defend the freedom of speech and printed protest of young revolutionaries on the campus. His departure from the University of Texas to spend a Guggenheim Fellowship in Germany was connected with a bitter argument with the university authorities over his financial support of a radical student newspaper which had been banned and had gone underground. Quarrels with his colleagues J. T. Patterson and T. S. Painter exacerbated the situation, already rendered difficult by Muller's role as host to two visiting Soviet geneticists, I. J. Agol and S. Levit. Quite a startling outgrowth of these troublous times in Muller's life was his address to the American Eugenics Society in New York City on 23 August 1932. The title of his address was "The Dominance of Economics over Eugenics." In what was an unusually brilliant style, even for Muller, he directed wit as well as keen analysis at the prevailing philosophy of the American Eugenics Society, and virtually demolished it with a single blow. Graybeards left the hall quivering and shaken, never to recover their poise. Eugenics itself, under the subsequent direction of Frederick Osborn, took a healthier direction toward realization of the importance of the social environment and awareness of the dangers of arbitrary or totalitarian controls over human reproduction.

About a year later, in the early summer of 1933, both Muller and I were for some time in the genetics laboratory of N. W. Timoféeff-Ressovsky at the Hirnforschung (Brain Research) Institute, a division of the Kaiser-Wilhelm Institute located in northeastern Berlin. Muller's endorsement had enabled me to secure a postdoctoral fellowship abroad, one of the coveted National Research Council fellowships established by the Rockefeller Foundation and doubly welcome during the years of the Great Depression. Muller had already made up his mind to go to Moscow, to remain indefinitely at the Institute of Genetics there as a senior investigator. The end of intellectual freedom in Germany was by then fully evident. Almost daily Nazis invaded the institute to take away persons of whom they were suspicious. Severe beatings and indefinite disappearance were not uncommon. Under these circumstances I had seen the first burning of the books, when volumes written by Jewish authors and collected from bookstores and libraries were piled up in the Opernplatz one evening and set afire to the accompaniment of inflamed harangues by Nazi speakers. I argued with Muller that in essence all totalitarian regimes were similar, for all of them suppressed freedom of speech and opinion, and all of them threatened the integrity of science. Muller was wholly unconvinced that this could ever be the case in a Marxist country, where the will of the people, so he asserted, was supreme. Muller's later disillusionment in the USSR was complete and bitter, although it was only after the rise of Trofim Lysenko to scientific dominance and the total endorsement of Lysenko's antigenetic ideas by Stalin and the Presidium that Muller realized the situation was hopeless. He then volunteered for service in a medical corps in Spain, in order to be able to leave Moscow at that time. It is in this context that Muller's essay on "The Meaning of Freedom" is to be read.

In this respect, Muller's opinions closely parallel those of Lyman Bryson, whose own book *Science and Freedom* (1947) it is doubtful that Muller ever read. Bryson, like Muller, would measure freedom on the basis of the total choices available to the individual. "It is clear," he wrote, "that men are not free to do things they have never heard of." Thus science, by affording novel choices, broadens the horizons of human freedom and makes more explicit and more certain the consequences of social and individual choices. Will increasing knowledge, then, make liberty more difficult? Bryson said no, not necessarily, for the possibility of freedom lies in the knowledge of consequences and the determination of action *by consent rather than by compulsion.* Muller, too, regarded the "limitations and disharmonies of their own inner natures" as setting limits to men's freedom even more than restrictive outer circumstances, and he pointed to the *joint* exercise of freedom as the appropriate humanistic goal.

Muller's devotion was unwavering to the ideal of peace between nations and their ultimate union in a single world community governed by world law. His essay "In Search of Peace" is very short, but it says a great deal. In his own actions, Muller exemplified what he praised as the ideal. At the beginning of the essay entitled "Man's Future Birthright," Muller refers so modestly to the Einstein-Russell manifesto that led to the establishment of the Pugwash Conferences on Science and World Affairs that one might scarcely recognize that he was one of the eleven original signers of the document. Muller attended the first and third of these conferences but rather soon dropped from active participation. It was not for lack of sympathy, but seemingly because he had little patience for those long and difficult discussions of the technical aspects of disarmament and political settlement that necessarily followed an appeal to the use of reason in international discussions and the clarion call to scientists everywhere to make a possible contribution to peace. The Pugwash Conferences, if they accomplished nothing else, will occupy an important place in history as a training ground in international negotiations of such persons as Jerome Wiesner and Henry Kissinger. Nevertheless, they often seemed at the time to have accomplished very little. Muller's devotion to his ideal of the scientific world society could be better exercised than in the tedium of seemingly endless negotiations or the acrimony of confrontation. His writings show best what he felt he had to contribute to the world on these matters.

As secretary, or rapporteur, of the Committee on the Biological Effects of Atomic Radiation which was set up by the National Academy of Sciences with funding from the Rockefeller Foundation, I had many opportunities to see Muller, in the circle of his peers, discussing the technical aspects of the danger from ionizing radiation, whether from nuclear weapons tests, from unanticipated explosions or leakages from nuclear power plants, or from the indiscriminate medical uses of high-energy radiations without proper safeguards. The substance of his views is well put in the essay on "The Radiation Danger," which needs to be read in context with the reports (1956 and 1960) of the above-named committee, if one is to see to what a great extent Muller's views influenced them.

The first of those reports was the first official statement urging caution in the matter and was soon followed by others backing up the conclusions reached. Those later reports, by the British Medical Research Council, by the World Health Organization (on whose committee Muller also served), and by the United Nations Scientific Committee on the Effects of Radiation on Man, have jointly been responsible for the far more cautious approach evident today in regard to medical and industrial exposures to radiation. Together with the Pugwash Conferences in 1960 and later, they also paved the way for the Atmospheric Nuclear Weapons Test Ban. Muller's role in pointing out the serious consequences of genetic exposures of the world's populations to X rays and other ionizing radiations was a leading one. At this date, one still wonders whether the Atomic Energy Commission or the State Department banned the attendance of Muller as an American delegate to the first Atoms for Peace Conference in Geneva, in 1955, for his past Marxian views and his residence in Russia or, rather, because of his passionate and persistent attacks on medical practice and atomic energy standards that permitted a dangerously high genetic dose of radiation to be inflicted on the American public. Now, in the eyes of all the world, Muller stands justified. The principles for which he contended are accepted both by his fellow scientists and the atomic energy authorities.

The most controversial of Muller's views are set forth in the three final essays in the volume. Will he in time be justified in these matters, too? I think so—not, perhaps, in the details of methods recommended nor without extreme difficulty in attaining the goals of eugenic improvement (of which more hereafter); but certainly so in respect to the growing concern of peoples everywhere for the genetic health of their populations, for the evolutionary betterment of the species, and for the increased expectation that, through some social regulation of reproduction, every child may be born physically and mentally sound and able to make the most of an improved environment. Muller has argued repeatedly for the practice of artificial insemination, which offers the possibility of eugenic selection through choice of the male donor. With rapid establishment of sperm banks in the United States—at least three new ones were created in 1972—the practice is certainly gaining currency. There seem to be no legal impediments to a wider use of artificial insemination except as respects the ultimate status of the child. Courts have held that a child so produced is legally a bastard, even though the husband of the mother may have given consent; and the inheritance of property has been rendered questionable. Psychologically, as many opponents of any extension of the practice of artificial insemination are quick to point out, there may be an unfortunate reaction on the part of the excluded husband who, though at first consenting, may subsequently come to feel that the child is none of his. The child therefore may become a basis of estrangement between husband and wife. The feeling is no doubt illogical, but it must be admitted that in sexual matters people are often not very logical.

From the eugenic standpoint, two arguments may be made against widening the practice of artificial insemination. The first, which time may correct, is that little or no genetic examination is made at the present time of the significant aspects of the genotype of the sperm donor. We may assume that, since most of us carry an average of four to eight or more detrimental recessive genes, individual sperm donors cannot be guaranteed to be free of them. Yet there is at present no systematic testing, such as could be made, to determine the possible presence of even those genes which might, by known methods, be detected. Already, according to a personal statement made by D. Y.-Y. Hsia, an authority on the subject, as many as one hundred serious inborn errors of metabolism or chromosome defects could be distinguished in the carrier by biochemical or cytogenetic screening. Such methods should unquestionably be made a preliminary to the preservation of spermatozoa in the sperm banks. The rule of practice ought, then, not to exclude the use of semen from a known carrier of a recessive genetic defect, for it is highly likely that all males are such. No, instead the procedure should require an application of similar biochemical and cytological screening to the prospective mother, so that any combination of ovum and sperm from persons who carry the same genetic defect might be avoided. The expense of such screening, at least at present, is not small; but then many of the genetic defects are exceedingly rare. Testing might be limited to those defects which are relatively common or which are exceedingly damaging to the future lives of the infants.

Both the legal-psychological and the genetic objections to artificial insemination can be avoided by a logical extension of the advocated practice of artificial insemination to that of artificial implantation of embryos produced by selection of ova as well as spermatozoa. The technical advances needed for this development in controlled reproduction are at the present date already at hand. R. G. Edwards and his colleagues at the Physiological Laboratory of Cambridge University, England, have successfully removed ova from female volunteers whose oviducts are blocked. After treatment with female sex hormones to induce superovulation, considerable numbers of ripe ova may be obtained by means of a relatively minor surgical operation. These ova have been fertilized in the laboratory with sperms from the husband of each respective ovum donor, and the embryos have been reared successfully to the blastocyst stage, when implantation in the uterus normally occurs, preparatory to formation of a placenta. Under the laboratory conditions, it is possible to observe the development of the embryo critically and to discard any which show signs of abnormality. It is even feasible to remove a sufficient number of cells from an embryo, without damaging it, to grow a cell culture that may be examined biochemically and cytologically for any indication of detrimental genetic traits or chromosome abnormalities, respectively. Such embryos may also be tested for their sex. In case both parents are carriers of the same recessive genetic defect, and there is consequently a probability of one in four that a given embryo may be homozygous for the defect, that eventuality too may be tested, and the embryo discarded if found to be doomed to serious defect. Atlhough Edwards and his group are moving only toward making it possible for an otherwise infertile couple to have offspring, it is quite clear that the technique, once established, might have wider application. That is, the implanted embryo might be chosen from selected donors of gametes other than either the husband or the wife. In this case a legal adoption would be necessary to establish the rights of the child, but *prenatal* adoption has much to recommend it over *postnatal* adoption. Not only could the genetic quality of the sperm and egg donors be selected and screened by tests, on both sides instead of the single selection of the sperm donor which Muller advocated, but in addition the biological and psychological experience of pregnancy jointly shared by both father and mother from the beginning should truly lead to a deeper and more intimate belonging of the child to the adopting parents. The only too frequent quarrels between a biological mother and adopting parents, in case the former changes her mind after submitting her child for adoption, could also be avoided. Although the concept of prenatal adoption seems very strange to many persons today, it has much to recommend it in preference to either artificial insemination, on the one hand, or postnatal adoption, on the other.

In recent years experimentation with the implantation of embryos in domestic animals has advanced to a high state of the art.

Cattle embryos have been shipped from continent to continent in the uterus of a rabbit, in which they can survive for some days, and have then been successfully implanted in cows of another breed, brought to full term, and delivered normally. The calves were in no way injured in the process. Also, pig embryos produced in Canada on a government farm were flown to Great Britain in a tube of nutrient fluid, implanted in a sow of a different breed, and successfully delivered. These experiments reveal that no biological damage is likely to be inflicted by the implantation of a human embryo in a foster-mother. On the other hand, the selection that can be exercised if embryos are reared in the laboratory to the implantation stage can be used to avoid any production of children with recessive defects, even when both true parents are carriers of the same recessive defect or when one parent is a carrier of a dominant defect, provided the appropriate biochemical and cytological tests are made. In the case of women over the age of 35 or 40 years, when the probability of the chromosome accident that produces mongolism (Down's syndrome) increases greatly over the probability in younger mothers, production of the embryo in the laboratory and a chromosomal diagnosis prior to implantation could do away with the majority of mongol defectives in our population, who now cost society, it is estimated, as much as $1.7 billion annually for care. Similarly, the diagnosis of Tay-Sachs or sickle cell diseases, when both parents are known to be carriers, can be made in the embryo prior to implantation, which might be a less objectionable practice than the wait to diagnose the condition when the fetus is 10 to 20 weeks old and then to abort it.

The second argument against the widening use of artificial insemination, or similarly against the introduction of embryo implants, is that they would have little if any perceptible effect upon the composition of the human gene pool. To this it may be answered, quite simply, that even the reduction in frequency of a single genetic defect such as Down's syndrome, Tay-Sachs disease, or sickle cell disease will benefit the population at large through the great reduction of the social cost of such defects, but especially that it will benefit the prospective parents who can then expect relief from the crippling fear of producing a defective child and yet fulfill the normal human desire to become parents. It is clear that the first introduction and application of these practices will be quite voluntary, in response to one of the most deepseated of human motives, namely, the desire, so very keen among the thwarted childless persons in our midst, to experience the joy of parenthood.

It has been pointed out by several geneticists, however, that to compensate for a genetically defective embryo by substituting a sound one derived from the same parents will tend to increase the frequency in the population of the recessive gene responsible for the defect. This follows from the fact that among the non-affected offspring of two heterozygous carriers, the expected proportion of carriers to noncarriers is two to one. However, if such parents were to choose for implantation an embryo derived from different sperm and egg donors, tested and certified to be free of the detrimental gene, they can actually contribute to a reduction of the frequency of that gene in the gene pool of the population, since they will not risk passing on their own genes at all.

We speak, of course, only of genetic disorders that are excessively severe and by present methods uncurable. To attempt to eliminate from the population all minor defects readily treatable by drug, diet, or prosthetic device would be impracticable. On the other hand, what can be said about the prospect for an actual improvement in the genetic stock of the population, about which Muller was even more concerned than about the elimination of severe defects? After earlier considerations of the problem, when he enumerated a rather large number of desirable goals for the selective breeding of mankind, Muller in his later essays on the subject, which are well represented by the final essay in this volume, reduced the goals to two characteristics, and no more: *cooperativeness* and *intelligence*. It is of exceedingly great interest that these two goals are identical with those which Charles Darwin had named, a century ago, in *The Descent of Man* (1871) as the paramount qualities that make man supreme among living beings. Darwin identified the two most significant human characteristics as being "sympathy," underlying mutual aid, and "reason," which enables man to learn by experience and to exercise foresight. The difference in vocabulary should not conceal from us the true identity of Darwin's selection of paramount characteristics and Muller's choice of the goals of human evolutionary progress.

The difficulty, which has not been solved during the past century, is to define these qualities in objective, measurable terms and to identify their genetic patterns of inheritance. Both of them are highly modifiable by experience and education. Each of them is culture-bound. Even for intelligence, for which tests have now existed almost throughout the twentieth century, there is no perfect—even partially certain—way of delimiting the genetic and the environmental components. The inheritance of intelligence, as measured by IQ tests, was of great interest to Muller throughout his life. He carried through one of the earliest studies of the intelligence of a pair of monozygous twins separated early in life. What was quite clear to Muller and to other students of the statistical variation of the IQ is that if you reduce nearly to elimination the variance in the genetic component, the remaining variance, attributable to the environmental component, approaches 100 percent. The separated identical twins clearly exemplified this axiom. But the variance that remains is also greatly reduced in absolute amount: to 20 IQ points in one pair of twins studied later by H. H. Newman, to less than that in almost all other such twin pairs on record. The average difference between separated monozygous cotwins is only about 7 IQ points, or scarcely more than the difference exhibited by one and the same person taking the test at different times, under different physical conditions. Conversely, the more uniform the environment is made, the greater the proportion of remaining variance which must be attributed to hereditary differences. Again, it approaches 100 percent. But that difference is far less than the existing range of IQ's in the population.

As for sympathy and cooperativeness, the problem at present is insoluble. One may feel strongly that there is a hereditary basis for this complex trait, somewhere in the hidden recesses of the brain or the appropriate balance of the adrenal hormones. Nevertheless, there is little way of measuring the hereditary component of cooperativeness exactly, or even of concluding positively that one is present at all. We cannot speak of it as definitely dominant or recessive, or as a single rather than a polygenic characteristic. Under these circumstances it seems futile to suggest that selection might enhance it. About all we can be sure of is that however and to whatever degree it is heritable, its modifiability by nurture within the family, education in the schools, and friendly or hostile reactions in work and recreation is so great that any genetic predisposition is heavily masked by the effects of the social environment.

Muller was of course well aware of these difficulties. In his final essay he has emphasized that all forms of discrimination against the development and education of children belonging to different social classes or races must be eliminated, if we are ever to see more clearly than now what differences in genetic heritage are in fact present. He also admits that selection, during an indefinite future, will have to be based upon empirical, "performance" criteria rather than genetic analysis. Surely the delay of some decades—even a generation or more—which he thinks ought to elapse before any use is made of donor sperms would, especially if applied to the ova too, greatly assist parents to make a pragmatic choice of the available germ lines. The self-nominated elite of "prime-movers" Muller advocated in order to get the selection program started poses obvious dangers. We must especially heed his warning against the egotists and paranoids who may wish to enter the ranks of prime-movers, supporters, and donors in order to seek the preservation of their own germinal material. These perils might be in great measure avoided if a sufficient delay in the actual use of stored germ cells permitted the sterner and more impartial judgments of a future century.

Muller was indeed fully aware of the need to perfect the technology of storage of both male and female germ cells. He mentioned the most recent advances in technology made in 1966 and would have been gratified to see what is already possible in 1973, six years after his death. Not only mouse ova, but even mouse embryos at the blastocyst stage have been frozen, held for days at the temperature of liquid nitrogen, and then thawed out undamaged. After being implanted in a foster mother, normal young mice were born. The skepticism of many persons, including probably a majority of the geneticists of his own generation, about Muller's proposals will not prevent the realization in due time of what he foresaw, if a few bold pioneers pursue the course he mapped out.

What genetic course will man steer? Although we cannot guess the outcome, the world already owes a lasting debt to Muller for his forthrightness in posing the issues. From Darwin's *Descent of Man* to Muller's last essay on the future of human genetics is just less than a century. What will the next century bring forth in this arena of man's conflicting hopes and desires? Once, in considerable doubt, I asked the question: "But can the geneticist breed wisdom, or integrity, or even simple humanity?" The answer may be, as I wrote some years later: "We cannot turn the clock back. We cannot regain the Garden of Eden or recapture our lost innocence. From now on we are responsible for the welfare of all living things, and what we do will mold or shatter our own heart's desire." That conclusion typifies the lifelong quest of Hermann Joseph Muller, who never lost faith that we could meet in wisdom and humility the challenge of exercising the greatest power vouchsafed to man, in order to create eventually a better race of men to succeed our own halting footsteps.

Bentley Glass

Man's Future Birthright

Possible Advances of the Next Hundred Years: A Biologist's View

THE kind of world our great-great-grandchildren will live in depends on whether there will in the meantime be a major war, and in case there is no war whether a democratic or an authoritarian system gains the ascendancy. Assuming that men succeed in evading both war and dictatorship, the kind of world they will make for themselves depends upon whether their reason will so far overcome their fanaticism as to persuade them to curb their numbers adequately. Provided that the answer to this question also is a favorable one, the amount of their progress depends on the thoroughness with which their reason triumphs: that is, on the extent to which men in general come to adopt the scientific attitude and to put it into practice in their daily living and in the organization of a world society.

So turbulent is today's situation and so devious and variable the conflicting currents, that only the rash or the ignorant would pretend to answer any of these four questions with assurance. Certainly, however, there is a chance for reason and freedom to prevail, if we work for them with all our great new tools of communication and enlightenment, of human organization, of industry, and of scientific research. For the purpose of the present discussion we shall assume that these four bridges will be crossed with reasonable success. We then find, in trying to forecast the advances to be made, that despite the enormous uncertainties that confront us in this proverbially unpredictable field we are nevertheless on somewhat firmer ground than in trying to prognosticate the outcome of those reactions of men that will decide the first four questions.

Assuming, then, that social conditions allow science and technology to continue to follow that curve of accelerating ascent which recent decades have set the pattern for, there is no doubt that long before 2057 nuclear as well as solar energy will have been harnessed to space flight and that the minor planets and major satellites of our solar system will have been visited. More ambitious plans will be afoot for expeditions, requiring crews that live successive generations en route, to planets of the nearer stars. These are not, however, means of relieving population surpluses.

On our earth, the taming not only of the great jungle areas but also of the deserts and the oceans will be well under way. These regions will be used to furnish food and other materials, as well as living space, although here again a sufficiency can be had only by restricting the increase of population. Granting this restriction, however, energy from atomic fission and probably also from fusion, together with automation, will bring within reach of the average citizen of the world at that time a standard of living higher than that of the average American at present. Moreover, he will have access to much richer and more satisfying leisure and to far more enlightening and inspiring education.

Great as the conquests over lifeless matter may be, those having to do with living things will be just as startling. My primary concern in this discussion lies in this field. The means of overcoming hostile microorganisms and parasites within the body will have been so far improved as virtually to eliminate ailments caused by them. As for the so-called degenerative diseases of advancing age —the cancers, the disabilities of heart and circulatory systems and nervous system, and the rheumatic conditions—methods of prevention and of treatment will certainly have been found that will considerably delay their onset and progression and reduce their incidence and severity. Nevertheless, the heart and blood vessels as well as the brain will still be subject to enfeeblement with advancing age, and it seems unlikely that means will yet have been found for radically increasing the span of life. Moreover, cancers will continue to arise, and it is probable that most of them, when not discovered so early as to enable their complete eradication by surgery, will still be incurable.

The organ concerning which the greatest revelations of the next hundred years are likely to be made is the brain. Proceeding beyond experiments on animals, investigators will undoubtedly probe into their own brains to find the relations of conscious to material processes. They will analyze these results in terms of other findings made on individual cells and groups of cells of the nervous system. The conclusions thus reached will inevitably usher in a series of revolutions in psychology, psychiatry, and philosophy.

Although, at present, replacement of parts of the body by transfer from other individuals is unsuccessful because of the tissue antagonisms that finally arise, modern discoveries hint that it may be found possible, as by preparing the individuals at an early age, to overcome such antagonisms. Such an achievement might have great possibilities not merely for replacing injured or ailing parts but also naturally defective or aged ones. Beyond this, the whole field of regeneration of members and of guiding the development of embryonic and premature stages may disclose ways of controlling growth and the differentiation of parts to enhance the individual in given respects.

The coming hundred years will see important advances in the control of human reproduction. In addition to bringing the reproductive cycle under regulation, it will probably become possible to prescribe the sex of the child and to produce at will twins, either identical or fraternal, or still more multiple births. The prevention of overpopulation, that is one of our assumptions, can occur only through the widespread acceptance of the philosophy that the number of offspring to be produced should be restricted for the good of those offspring themselves. With this more ethical attitude regarding reproduction, it will also be regarded as a social obligation to bring into the world children as favorably equipped by nature as possible, rather than children who closely mirror their parents' peculiarities.

In the service of this new morality, foster pregnancy, already possible, will be readily achieved and widely welcomed, in addition to natural pregnancy. This will provide the opportunity of bearing a child resulting from the union, under the microscope, of reproductive cells one or both of which were derived from persons who exemplified the considered ideals of the foster parents.

These reproductive cells will preferably be derived from persons long deceased to permit a better perspective to be reached on their worth, one relatively free from personal pressures and prejudices. For this purpose, banks of deep-frozen reproductive cells will be maintained, as well as multiplying cultures of them. This procedure will make the most precious genetic heritage of all humanity available for nurturing into childhood and adulthood by the many devotees of human progress who would cherish such an undertaking.

Even more predictability concerning the nature of the progeny will be attainable, when desired, by a kind of parthenogenesis, or reproduction without fertilization. Whereas offspring ordinarily have their hereditary material picked in a random way from two different parents, in this case the offspring obtains his hereditary equipment entirely from one individual, with whom he is as identical genetically as if he were his identical twin. This will be accomplished by extracting the nucleus from a human egg and inserting in its place an entire nucleus obtained from a cell of some preexisting person, chosen on the evidence of the life he or she had led and his or her tried potentialities.

Fortunately, men will in all probability have joined into one world community before these techniques come into widespread use. For if the people of one nation were to apply them intelligently and extensively even a few decades before the rest of the world did so, they would be able soon afterwards to rise to such a higher level of capability as to make them virtually invincible. The world cannot afford to have separate nations putting up their separate genetic sputniks!

Our insight into the chemical basis of bodily and cellular operations will during the same hundred years be enormously deepened and broadened. We shall certainly have learned how to create living things of the simplest types and shall be advancing up the scale of microorganisms.

Much will also be known to us of the principles of operation of the so-called code presented by our own hereditary material. According to some calculations, the message in this code, if spelled out, would be likely to fill about a hundred volumes, each of the size of the *Unabridged Webster's International Dictionary* and with equally fine print. In this connection, we will realize the importance of not allowing radiation or other harmful influences to damage this precious heritage. On the contrary, we will increasingly take advantage of the rare favorable changes that have arisen in it naturally, by allowing those that have definitely proved their worth to be transmitted to increasing numbers of descendants.

Our measure of worth in regard to genetic endowment must be: whatever tends to make man's nature nobler, more capable, more harmonious, more sympathetic, happier, and more beautiful. Granted the triumph of peace, freedom, and rationality, this ideal will within a century be a widely accepted one for human effort. Unlimited progress, continuing evolution, may be possible in these directions.

Science Fiction as an Escape

SCIENCE fiction is often denounced as an escape by pedants, scowling down their blue noses. Of course it is an escape; why deny it? The intended accusation contains an unintended compliment. For unless we are content to side with pessimists who define life as "just one damn thing after another," we must admit that we are forever engaged in trying to escape into a better situation. And why should we not be doing so?

In Defense of Escapes

What is the object of all art and, for that matter, of all science and of all attempts at better living, if it is not to emancipate ourselves more and more? Why try to build "ever more stately mansions," unless to escape to higher freedoms? What then is freedom itself but an eternal escaping and, thereby, an eternal attaining? Would we relinquish that "divine discontent" that keeps pressing us onward?

We must admit, however, that while some routes provide genuine means of escape, others—such as an addiction to opium—are mistaken, in that to follow them is ultimately self-defeating. In which class is science fiction? It can be of either type, or something in between, according to how much it helps or hinders us in facing the real world and meeting its challenges.

That real world is increasingly seen to be not the tidy little garden of our race's childhood, but the extraordinary, extravagant universe descried by the eyes of science. Moreover, the format of that part of this real world that immediately surrounds us has already been considerably reshaped by the tools provided by science. And by means of these and better tools yet to be fashioned, this reshaping must be continued apace, with the aim of enabling us to avoid the ever deeper pitfalls and to attain the ever higher summits, which we see emerging from the mists ahead.

It is inevitable and necessary for this change in our outlook and methods to be expressed in all aspects of our life, including our art. If our art and, especially, our more intellectual fabrications do not explore the relations and contingencies implicit in the greater world into which we are forcing our way and do not reflect the hopes and fears based on these appraisals, then that art is a dead pretense, a disappearing vestige. But man will not live without art. In a scientific age he will therefore have science fiction.

In such an age, men as individuals have incomparably greater capabilities of experiencing and of accomplishing than were ever before at their disposal, while men as a group have powers that utterly belittle those of the strutting gods and heroes of former times. Yet men's range of vision and their appreciation of possibilities have at the same time run so far beyond what they can directly experience in their own lifetimes, and even beyond what is presently accessible to them as a group, that they feel much more need than in the past of gaining vicarious experiences that range ever wider from their immediate base. Thus they grope to bring home into their relatively limited apperceptions, through fiction, as vivid a visualization as they can—admittedly retouched—of the shapes glimpsed in the dimly lit distances of space, time, and circumstance.

Although aware that the particulars are invented, men are thereby enabled to attain a truer perspective on their individual affairs and also to be more useful in reorienting the joint endeavors of their group. In other words, as man-the-group progresses in framing his thoughts and acts with reference to things and processes of increasingly longer range, men as individuals, must be given the opportunity to share in this expansion of view and motivation and to feel as directly as possible the thrills that their larger role can bring. The enhanced sense of participation thus engendered will serve in sustaining, stimulating, and to some degree in guiding them, both in their separate and in their joint efforts.

It hardly needs to be pointed out that a prime requirement for science fiction, if it is to fulfill the function just formulated, is that it be entertaining. This is by no means synonymous with jollity or with having a happy ending, for tragedy is often more deeply strengthening than victory. But the story should be one that absorbs and convinces us and that at the same time affords us relief from our daily doings by taking us via the narrative, not the didactic, route satisfactorily beyond our accustomed horizons. Properly to meet this combination of conditions is a job that demands an extremely high order of abilities. These must include not only a working understanding of the major principles and possibilities of present-day natural science and technology, in the diverse lines relevant to the theme dealt with, but also a rounded insight into human relations and feelings, a fertile but well-controlled imagination, and the exacting skills of a writer. It is small wonder that, with our modern specializations of training, few writers of today and still fewer of yesterday have measured up to the standards thus set.

Spurious Escapes

In addition to its inherent difficulties, science fiction is today confronted with unusually corrupting influences. The wave of popularity conferred on it since 1945 by the development of nuclear energy and of rocket propulsion has been of very dubious benefit to its healthy growth. It has provided a multitude of miseducated and undiscriminating readers whose demands have created a flood of "literature" adapted to their tastes, and often serving to mislead them further. Within this deluge it is difficult to discover the relatively rare items of genuine science fiction, since the existing journals, anthologies, and book series purporting to dispense science fiction tend to be as unselective as the general public to whom they cater. Thus the ostensible science fiction story of today, when taken in relation to the level of science contemporary with it, is *on the average* on a considerably lower plane than that of fifteen or twenty years ago. The absolute number of good stories is doubtless greater than before; but the total of all, good and bad, has become so inflated that the good ones are now harder to find.

It is to be expected that themes once highly original—as space flight and the meeting of men with intelligent creatures of other worlds were in H. G. Wells's superb *First Men in the Moon,* or time travel in his *Time Machine* and Mark Twain's *Connecticut Yankee in King Arthur's Court*—will become trite and tawdry when reiterated by commonplace hack writers. However, we may well admit that there is a function for these latter-day disguised Westerns, in satisfying the cravings of juveniles in a manner designed to make them somewhat more forward-looking than the Westerns would, although this is hardly science fiction according to the standards set forth above.

It would be over-puristic to complain at the unintentional lapses from scientific fidelity that the best of writers will occasionally fall into. Nor need we begrudge such writers the ruses whereby, in intentionally violating some one or two scientific principles, they provide the opportunity for the development of a story that for the most part rings true and conveys notable meanings. In fact, the three stories just mentioned are cases in point, for both the antigravity device of Well's moon story and the time travel of the other two stories are incompatible with the known workings of nature. But, after having been made to seem plausible by clever circumlocution (or, in Mark Twain's case, by the total avoidance of analysis), these tricks are merely used to shift the scenes, which thereafter proceed by their own logic. In such a case we can grant the writer his license and conveniently close our eyes to the maneuver.

It is different, however, when the main theme of a story is centered about an impossibility, and more especially so when the fallacy is one that is prevalent, is intended to be believed, and is currently the source of important popular misconceptions. The favorite time-travel stories and some other "impossibilist" fiction of today hardly fit into this class. For although their plots often are centered about the contradictions they engender, most of them may be dismissed as frank fairy tales not intended to be credible.

The really serious element of antiscience that is the most objectionable in much of the so-called science fiction of today is its dualism: its failure to recognize the dependence of mind on body. In this lapse it is representative of the current recrudescence of old-time theology and of animistic superstition in general, with all their associated spiritistic and mystical beliefs.

Superstition Posing as Science

This revivalist zeal, which during the past twelve years has seized upon public officials and the mass molders of opinion and, through them, upon the general public, seems to have arisen as a phase of the cold war. Taking advantage of the general revulsion against the methods of the Communists, and also of the fact that the Communists have repudiated most of the dogmas of old-fashioned theologies, some of the supporters of these dogmas have been trying to get them recognized as constituting the heart of the Western position. In so doing they ignore the fact that the Communists, with their dialectical materialism, Lysenkoism, doctrine of inevitable class struggle, etc., are in their own not-so-new way just as mystical and dogmatic as they, the reactionaries, are; and that what is most objectionable in the Communist position is precisely that principle of authoritarianism and of intellectual and moral slavery in which the Communists not merely agree with the reactionaries but actually go them one better.

It is time for this situation to become more generally realized. It will then become evident that to attach the appellation "science fiction" to fables that merely represent outgrowths of the ancient dualistic superstition of all primitive men is a flagrant reversal of the meaning of words, similar to the reversals so frequently practiced, for their own ulterior reasons, by the Communists themselves. It is these pseudo-science stories that offer men escape in the pernicious sense: that of an opiate which withdraws them from reality.

The doctrine of dualism is so deeply rooted in human traditions, it has been held so long in awe and has flourished in so many forms, it still permeates so much of education (especially education of the very young) and of everyday language and practices, that it is to be found among us on every side and in a multitude of manifestations, many of which fail to be recognized as such. It is of course basic to all notions of the supernatural, whether in their more philosophical or their more primitive forms. For the supernatural always involves the workings of bodiless spirits, or at least of disembodied ones. These mysterious or magical entities can dominate, interfere with, and circumvent physical forces, which lie on a lower level, as it were. Thus they can create, annihilate, transform, or "possess" material objects, as well as other minds, and can on occasion traverse space and time almost at will. All these "phenomena" are found at work in much of the so-called science fiction of today.

In the genre of fiction inappropriately designated nowadays as "fantasy," the old-time ghosts, jinn, demons, and even gods have been resuscitated. They stalk about in the open or lurk in shadows, as the case may be; but they practice their ancient sorcery without apology in modern and not-so-modern settings. However, in the genre posing as science fiction these spirits and their workings, although often dealt with, are usually disguised by couching them in pretentious sham-scientific terms and by expositions calculated to deceive the unsophisticated into regarding these phenomena as physically and biologically plausible. Among such treatments the more nearly genuine science fiction often finds itself intercalated.

In these ways the reader tends to get the impression that the scientifically conceivable inventions and the disguised spiritism are akin and similarly reasonable. This impression is frequently reinforced by editorial argumentation and by allegedly factual articles. The wiping-out of the meaning of science is carried even further by a class of publications that makes it a point to offer stories of all three types interspersed: the frank fairy tales, the "scientifically" camouflaged ones, and the true science fiction, together with all grades and mixtures between them. The same practices are followed by many movie, radio, and TV presentations.

Bordering all branches of science there is of course a "lunatic fringe" of wishful thinkers to be found defending some bogus cancer cure, mysterious radiation effect, or species of dualism. Among the latter should be classed postulates of cellular intelligence or memory, vital force, perfecting principle, cosmic purpose, extrasensory perception (ESP), telepathy, telekinesis, clairvoyance, psionics, dianetics, etc. These claims, occasionally finding their way into reputable scientific publications, are seldom taken seriously by the scientists in the relevant fields, but they afford a wealth of material for the popular type of "science fiction." The fact that many justifiable scientific hypotheses and genuine findings also have to go through a period in which they represent a minority view and must then contend with bitter opposition makes possible the pretense in any given case that the former are in a class with the latter. It thus becomes easy to hoodwink the wishful reader, lacking the background for judgment, into acceptance of the spurious "principle" allegedly so unjustly condemned; and the reading or viewing of a well-told story based upon it tends to seal his belief in it.

Those who contend that little harm is done by the spreading of such misconceptions among the uninitiated are losing sight of the urgent modern need for the dissemination of genuine scientific principles among the public at large, including their leaders, and of the extraordinary difficulty experienced by teachers and science popularizers alike in instilling more rational conceptions. Good science fiction, presented in publications that could be depended upon to avoid (or to label) the fraudulent, would go a long way toward meeting this need.

It may be objected that most of the examples cited above as representing unscientific ideas have never been categorically disproved, and that it is therefore unscientifically dogmatic to throw them out. A similar claim can be made by the man who believes himself to be Napoleon, and he can usually find some supporting evidence for his belief. In judging the validity of a conclusion we should take into account not merely the specific observations in question. These may have been subject to sensory errors, peculiarities of statistical fluctuation, and systematically misleading conditions of the innumerable kinds that can trick even careful students. We must also test the apparent result for its consistency with the whole body of previous knowledge.

Thus, in the field of mind-body relationships, all the evidence that has withstood searching tests joins in indicating that what we loosely term "thought processes" are indissolubly bound with the functioning of a most highly organized structure of a very exceptional kind: the brain, for which no parallel exists in other natural objects. It indicates further that the data for these thought processes are supplied by the impulses coming into the brain through the sensory nerves, and that the processes produce external effects of a "willed" kind only by initiating impulses in the outgoing nerves. Of course, we *may* postulate that the thought processes can nevertheless exist separately from the brain, or that the brain can selectively influence or be influenced by objects external to it—including other brains—in other functional ways than through its nerves. Similarly, we *may* postulate that the hidden side of the moon is inhabited by a race of human beings. However, when we make such an assumption we abandon scientific ways of thinking and take refuge in a kind of thing that is only infinitesimally possible.

It is only to be expected that unoriginal, uncritical minds, in trying to be fanciful, will fall back (wittingly or not) on the myths of their childhood. Refurbishing these with the frippery and the patter of today, they will offer them to others as novel inventions. Fortunately, however, this is not the road of "escape" for fertile imaginations. For them the possibilities that are consistent with the already known principles of natural science loom up so manifold and remarkable as to give limitless scope for the exercise of their constructive speculation.

Perhaps it is because I am a biologist that so much modern science fiction seems to me to be especially wide of the mark on the biological side. It is here of course that dualism is most rampant and that mysterious spirit forces take the place of neural processes. Not infrequently they are also made to take over the process of evolution. In the following passage from a story in a science fiction magazine of September 1957 we find both forms of spiritism at once.

"Nature has a way of providing the particular trait just at the time it is most needed. A good example is the way more male children are born during a war. There's no known explanation for something like that. But nature seems to know what is needed—and provides it."

"That sounds plausible," Saxton said, after a minute of consideration. "According to your theory, then, those savages possess an ability radically different from that of normal humans?"

"Not necessarily radically different," Wallace answered. "It would probably be a trait inherent in all of us, but not so evident, or fully developed. Or perhaps it has made its appearance before, in rare individuals, but not being a survival characteristic—where it appeared—it died. Something like telepathy, or poltergeism, or any of the other so-called wild talents."

To me, after nearly a half-century of struggle to make people aware of the correctness of Darwin's theory of evolution by the natural selection of fortuitously arising (i.e., unplanned) variations—usually of minute degree—and much time spent gathering evidence on this matter, such statements as the one above are ideologically dangerous. Closing off the way to one of the areas of science most important in the formation of our general conceptions of the nature of things, they reopen the door to superstition. It must be admitted that behind these manifestations in our fictional treatments there lies the flagrant inadequacy of the teaching of evolution in our schools, and that this in turn is an expression of continuing pressure from the orthodoxies. For this very reason, however, a science fiction free from such distortions is the more needed.

Spurious Evolution

Without some grasp of the long-known fundamentals of evolution and genetics, the science fictionist—even when not a dualist—is likely to fall into other prevalent errors and to relay them to his readers. One is the idea, originating in a distortion of some findings of my own, that radiation will produce a race of monsters. A whole school of "mutation" stories (including world's end stories), founded on this distortion, has doubtless provided one of the main bases for the current hysterical exaggeration of the genetic damage done by the fallout from nuclear test explosions, and has thereby led to unrealistic appraisals of international policy. This disclaimer is not, however, to be construed as a denial of the genetic damage, nor of the importance of taking it into account in all our dealings with radiation, whether in war or peace. The situation should serve as an object lesson on the dangers that lie in the oversimplification and sensationalizing of science.

Another important misconception of evolution that is being propagated by a host of science fictionists is that on visiting other habitable planets we may find that the living things which evolved there fit into our terrestrial classifications. Commonly the creatures are even supposed to include "humanoids" that stand on two legs, have hands with fingers (although perhaps four or six!), hair (green, perhaps), a combination mouth for the three different functions of eating, breathing, and making sounds, and all the rest of it, up to nasal and chin prominences, eyes (three perhaps) and ears in their "right" places, human facial expressions, human secondary sexual characters, and human love-making! Of course their biochemistry is essentially the same as ours, their foods are edible to us, and drugs affect them similarly. Few of their differences from *Homo sapiens*, other than those mentioned, are greater than those between our known races—with which indeed some stories allow them to cross.

It has recently become fashionable among the more "sophisticated" science fictionists to scoff at the so-called bug-eyed monsters (BEM) of earlier tales. It happens however that, despite the crudities of some of these creations, such as their comparative lack of integration and their often being a mere mosaic of the features of different animals known to us, they came far nearer to giving the reader a valid impression of the sort of things to be expected elsewhere than do the ostensibly conservative representations of virtually earth-type beings that many of these scoffers are offering us. Despite their occasional flaunting of technical biological terms to impress the reader, these writers need to go back to the primary grades in evolution study. Biological history no more repeats itself in detail at different places and times than human history does. And it is almost as absurd to expect to find humanoids or dinosaurs or insects or birds on other habitable planets as it would be to expect the creatures to be communicating in English or Chinese. Of course few of the readers object to this absurdity, because most of them get their ideas of the subject from their writers.

Perhaps to many physicists the disintegrating beams, "force fields," "hyper-space," "space warps," "faster-than-light speeds," "antigrav arrangements," "sub-nuclear forces," and "parallel universes" that have become the standard paraphernalia of much science fiction appear as unreasonable as the biological presumptions do to me. However, we may well leave the discussion of these matters to them. In any case, they are made use of (strange to say) more or less independently of the biological and anthropological assumptions.

Human Nature and the Social Order in Flux

What may be of greater interest to readers of this essay book is the science fictionists' treatment of human personalities and relationships. In this field it should be acknowledged that some progress has been made in recent years. For one thing, there has been, on the whole, an improvement in character portrayal, and in the delineation of the repressions and subconscious motivations that play so considerable a role in human life. Secondly, the heightened public awareness of the evils of dictatorship and authoritarianism, engendered both by our struggles against the tyrannical systems of other countries and by the excesses of repression that these struggles have sometimes, in reaction, given rise to in our own midst, has had some notable manifestations in the science fiction of the last decade or two. In contrast with the social puerility of most earlier science fiction (that of H. G. Wells excepted in this respect), it has now become good usage to give serious and even searching consideration to problems of this type.

Along with these developments, as an expression of the same tendencies, recent science fiction must be accorded high credit for being one of the most active forces in support of equal opportunities, good will, and cooperation among all human beings, regardless of their racial and national origins. Its writers have been practically unanimous in their adherence to the ideal of "one free world." In criticizing them in other ways, let us not overlook their distinguished record in regard to this vital concern of the present day.

In their treatment of other areas of psychology and sociology, however, modern science fictionists have for the most part been disappointing. Aside from their exposure of the ills of dictatorial regimes of already known types, extended by some rather obvious extrapolations and sometimes modified by automation, they have failed except in the rarest instances to explore boldly the enormous possibilities of the diverse changes that might take place in the course of generations in economic, political, and social relations, and in the structure of society in general. Certainly this would be a most formidable undertaking, worthy of the most brilliant and emancipated minds, especially when it is remembered that the impacts of changes in technology on the social structure must be weighed in any such treatment.

The contemporary American climate of smugness with regard to our forms of economy, politics, and sociology would not be conducive to the favorable reception of such efforts. Moreover, most present attempts of this kind can be expected to be primitive, floundering, and incapable of leading to definite conclusions. We should therefore be lenient in judging them, remembering that the tackling of this job is to be encouraged. There is no doubt that science fiction, with the literary license accorded to it in its depictions of human events of the future, as well as of conditions among fancied alien societies, has an opportunity to be of great service along these lines.

Together with the changes in human institutions, ways of doing things, social organization, and education that will take place in the future, and the correlative changes in the nature of the social problems, there are bound to be enormous changes in the psychology of the individual—including his motivations, hidden and overt—and his ideology. These too have been comparatively little touched upon by science fictionists. With a few notable exceptions, the man of the future and the "humanoid" of other worlds are depicted as largely the same in outlook and impulses as the man of today, although they may operate on a larger scale. Yet a glance at the findings of anthropologists shows that, despite the essential identities in the genetic bases that underlie human feelings and drives, they work out to give remarkably different attitudes and systems of conduct in different primitive human societies. How much more different, then, may they not become when subjected to the molding influences of civilizations more advanced than ours?

Here are real opportunities for the exercise of creative conception, not in the depiction of the people of static utopias of the kinds sometimes imagined in the nineteenth century, but in the redrawing of human urges in better adjustment to a progressive, cooperative striving for ever richer, greater social and individual lives. This development involves an increasing harmonization of the individual with the social, as well as the attainment of increasing harmony within each individual. No more will the science fiction hero have to seek escape by running away to new frontiers where he establishes his family and friends in virtual isolation from the oppressions of the larger society. No more will his inventions be pictured as purely individual products. Instead it will be shown how, in an advancing society, the individual will be given more opportunity to make significant and satisfying contributions to joint efforts of inspiring character and will in turn be the recipient of enhanced opportunities for worthwhile experiences.

This is a direction of narration and exposition far from sufficiently exploited. It could do much to counteract the pernicious fear and even hatred of science and scientists that modern misuses of advanced technologies have carried in their train. At the same time there will also continue to be an important role for narratives that show the consequences in societies in which the human readjustments have *not* sufficiently kept pace with the technical ones.

Challenging the Constellations of Genes

So few science fiction writers have a comprehension of biological fundamentals that they have virtually ignored the far-reaching potentialities of genetic change in the bases of human nature. If it is true that they have made disappointingly little use of their opportunities for depicting how much human nature, with its present genetic constitution, may be changed under the influence of the evolving cultural environment, it is even more true that they have neglected the profound role that genetic processes (both natural and more especially artificial) may play in the future evolution of man. The most they do in their occasional light brushing of this topic is to hold up to ridicule a form of statism in which, although marriages are arranged by a tyrannical eugenic dictatorship, people are still much the same as before, except perhaps for some stylization of features and submissiveness of disposition. In this area of science fiction, Victorian prudery and shortsightedness still reign as supreme as ever. In fact, the taboos today seem, despite our reforms in some matters of sex, to be far stronger than they were a third of a century ago, in the days of J. B. S. Haldane's *Daedalus,* or more than half a century ago, when G. B. Shaw declared in his *Man and Superman:* "Being cowards, we defeat natural selection under cover of philanthropy; being sluggards, we neglect artificial selection under cover of delicacy and morality."

In biology, so much water has flowed under the bridge in this last half century as to bring into view the possibility of biological techniques that could go far beyond those available when Shaw wrote, and that would at the same time be consistent with and conducive to the highest values in human relationships. If superstition did not lie in the way, this breakthrough could initiate a story of human advancement that would dwarf all previous evolution. This story, however, is one that is too large to be entered upon here. But it is one that science fictionists, if they are worth their salt, should be eager to develop in its manifold aspects. It should provide them with unexampled opportunities for leading the public, through convincingly told narrative of an unaccustomed kind, to the recognition of the inspiring possibilities here lying latent.

These possibilities may be brought into being if men will apply to the exploration and governance of their inner constellation of genes the same degree of intelligence, zeal, and industry as they are now using in their preparations to raise themselves towards the constellations in the sky. In the latter task, the most challenging external one, science fiction has admittedly been of great service in enlisting the enthusiastic support of the public. However, in the internal task, which is ultimately to be an even greater one, it has not yet even begun its operations.

To be worthy of its name and function, science fiction must be ahead of its time, leading people onward. It must jump gaps to give them views too daring to pass muster with the editors of less licensed publications. Yet it must keep the ground in sight in the sense of not contradicting established principles or gawking after the spirits of former times. For this, it must clean out the cobwebs from its own house and let in the light. It will then perceive ample panoramas, through the hills, jungles, and deserts of which to send its telephoto lenses for our benefit. We sorely need its speculations, but these speculations should be genuine. We have had enough of the phantasms dreamed up by our forefathers.

Life Forms to be Expected Elsewhere Than on Earth

IN this age of beginning space flight, many of our more alert students are avidly reading the literature that is commonly miscalled science fiction, and for this they cannot be blamed. As yet, it has provided them with almost their only available source of ideas concerning the possibilities of life on other worlds. Yet there is much in existing biology that bears on the interesting questions here involved. Let us review some of these matters, in order to lay a basis for giving our students better guidance in this legitimate quest of theirs than is currently to be found in the fabrications of romancers. We shall soon find that such a study leads through some of the liveliest problems of modern biology.

We may first recall an incident that happened soon after the great Polish astronomer Copernicus published his evidence, in the late 1500s, that the sun is far greater than the earth, that the earth and planets move around the sun, not the sun around the earth, and that stars are other suns ever so much farther away. This was in reality a view that had been originated by the Greek Aristarchus 2,000 years before Copernicus but that had been ridiculed as contrary to the religious teachings of the time and had been almost forgotten. In fact, Copernicus's great book was not published until just after his death, since he was afraid to have it appear while he was alive.

Shortly after it was published the Italian clergyman Giordano Bruno, convinced by this discovery, pointed out that there are probably many worlds other than ours, other planets belonging to stars other than our sun, on which living things exist, some of them intelligent beings. Chiefly because he refused to take back this view, Giordano Bruno was in the year 1600 burned at the stake by the Holy Roman Inquisition. It is reported that as he was led to the stake he said: "Perhaps you who pronounce this sentence against me do so with greater fear than I who hear it." If he said this, he was in reality accusing his accusers of attempting to suppress the search for truth because of their suspicion that they might themselves be shown to be wrong.

Planetary Origins

At any rate, the idea would not down, and it grew very prominent after the German philosopher Immanuel Kant and, independently, the French mathematician Marquis Pierre Simon de Laplace around 1800 proposed that the whole solar system of our sun and planets had originated from a great cloud of gas by its gradually falling together—condensing—as a result of the pull of gravitation on the particles. If this had happened in the case of our sun, it was realized, it might have happened similarly in the formation of millions of other stars also. So there might well be countless planets, and on many of these life might exist. However, during the early years of the present century, astronomers became very skeptical of this view of the origin of planets because of difficulties they encountered in explaining why part of the gas, in condensing, had got left so far out, as planets, supplied with such a high energy of rotation around the sun and around their own axis.

It was then proposed, instead, that the planets had originated by the sideswiping of one star by another one. It was rightly calculated, however, when the great distances between the stars and their speeds with respect to one another were reckoned with, that such an event would happen so very rarely as to allow hardly any stars to have acquired planets. But instead of concluding that this result made their hypothesis highly unlikely as the explanation of how our own sun's planets had arisen, they decided that our sun must be nearly unique in having planets. If this were true, life could hardly exist in our galaxy except on planets of a very few stars, and there was practically no hope left for any intelligent life to exist in our galaxy except on our earth.

However, during recent years new findings bearing on this matter have come to light and new calculations have been carried out on the way in which matter scattered in space must aggregate. All this new evidence indicates that the view proposed by Kant and by Laplace was correct in its essentials after all. Mechanisms have been worked out by which, in the condensation of gas and dust to form stars, not all of the material accumulates at one center. As Gerard P. Kuiper has pointed out, the evidence from double and multiple stars shows that, as expected, the gas more often condenses into two or more main centers that revolve about each other, and that there is a random distribution of sizes for these bodies, one of them frequently being much bigger than the others. In the limiting situation, which can be calculated to occur in at least 5 percent of cases, the great bulk of the material settles into one central star, or sun, while the rest is so small in amount that the other body or bodies would fall under the definition of planets and would not naturally support thermonuclear reactions of their own. Such planets could range in bulk from bodies much larger than Jupiter to those of sizes like the planets in our own solar system.

Although even the largest could not be seen telescopically by present methods, because of the great remoteness of other stars from us, nevertheless this conclusion has been backed by observations showing small perturbations in the motion of some stars, which prove that smaller dark bodies, much larger, however, than Jupiter, must accompany them. Bodies still smaller than this could not cause perturbations large enough to be detected by us. Nevertheless, all these points hang together and lead astronomers back again to the view that outside our own solar system there exists an enormous number of stars possessing families of planets.

Conditions for the Formation of Organic Compounds

But before a planet can support life, it must be supplied with the right chemical elements. It has been concluded that the earliest stars and their planets must have consisted of so little else than the very lightest elements, hydrogen and helium, as not to have allowed the formation of organic substances in sufficient amount for the maintenance of life. Only long afterwards, as these stars in their death throes formed somewhat heavier elements deep in their interiors, and then spewed their contents out into space, was it possible for this material to recondense. In this way it gave rise to later generations, as they are called, of stars and their planets. These were supplied, from their beginning, with more of the moderately heavy elements than the first stars had. Included here were not only carbon, oxygen, and nitrogen, but also, after more cycles, such elements as phosphorus, magnesium, and iron, all of which are necessary for life as we know it. Our own sun, it is thought, is a star of a third or even later generation, and any earlier generation than it would probably not have been provided with enough of these strategic heavier materials for the needs of active living.

Besides having suitable elements, a planet that supports life as we know it must have its temperature within a certain moderate range. Water constitutes the medium in the cell in which the chemical reactions that characterize life take place. Thus, while the operations of life are going on, the temperature must be above that of ice and below that of steam, and it must never rise high enough to destroy the essential organic compounds. In our own solar system, this condition would not allow life on a planet averaging much farther from the sun than Mars nor much nearer than Venus. The planet should also rotate on its axis at a fair velocity in order not to accumulate too much heat on one side. Moreover, its orbit must be sufficiently circular to keep it from ever getting too close to its sun. Finally, its sun, unlike many stars we know of must itself remain very stable in temperature over enormous periods.

In addition, any planet on which life as we know it exists must be of moderate mass. For if too small, like Mercury, its atmosphere and water would not be held down strongly enough by gravitation and would have evaporated off into space. And if too large, like Jupiter or Neptune, the planet would have prevented even its free hydrogen from evaporating. In that case chemical reactions favorable for life, even if started, would tend to be swamped out by the deep and all-pervading hydrogen quilt.

It has been estimated that even though all these conditions must hold at once there must be among the hundred billion or so stars of our galaxy several millions, at a low minimum reckoning, that would have one or more planets affording an appropriate setting in which living things might originate and sustain themselves. But how is it now thought that this origination could have taken place?

To understand this matter, let us first take into consideration the fact that even in stars like our sun that have a fair amount of the heavier elements, hydrogen is still greatly predominant. This must have been true likewise in the early stages of all our sun's planets, since they and the sun were derived from one common cloud of matter. The great excess of hydrogen would react with most of the other elements present that are necessary for life, such as oxygen, carbon, and nitrogen. As a result simple compounds would be formed of these other elements with the hydrogen, such as, for oxygen, water; for carbon, methane; and for nitrogen, ammonia. On a planet having a mass and temperature somewhat like the earth's these molecules are heavy enough to be held down within its crust, ocean, and atmosphere, while the free hydrogen, as we have noted already, is light enough soon to evaporate off into space.

Now the hydrogen compounds, or "hydrides," will be exposed to tiny bolts of high-intensity energy of several kinds. Those in the atmosphere will be subjected in greatest measure to ultra-violet light from the sun, and to a lesser extent to strokes of lightning, and both those in the atmosphere and in the crust will be subjected to a small but persistent bombardment with ionizing radiation that is similar in its action to X rays but is derived from cosmic rays, solar rays, and radioactive substances of the earth itself. All of these agents, brought to bear on the mixture of simple hydrides, have the effect of combining them together, so as to form larger, more complex molecules, in part oxidized. That is, organic compounds of diverse types are brought into being.

Thus it was shown by Stanley Miller, when he was working as a student of Harold Urey at the latter's University of Chicago laboratory, that when electric discharges are passed through a mixture of water vapor, methane, and ammonia, various amino acids (the building blocks of protein), and other organic molecules are produced. Moreover, Dr. Melvin Calvin of the University of California, on analyzing meteorites that had fallen to the earth from interplanetary space, found that some of them contained various organic compounds. Included among these were even some built in the form of rings and double rings composed of both nitrogen and carbon (and therefore, called heterocyclic). They were, in fact, closely akin to those structures, called purines and pyrimidines, which form the most distinctive part of the hereditary chemical core, the nucleic acid, of all living things. So we see that the necessary precursors of those types of organic molecules that play the most essential role in life, namely the amino acids, which are precursors to the protein, and the purines and the pyrimidines that are precursors to the nucleic acid, are both formed by natural processes that take place under primitive conditions.*

In the atmosphere of the primitive earth, before it contained free oxygen, the sun's active ultraviolet rays penetrated in much greater abundance than they do today. For the free oxygen of our present air, existing even at great heights, is converted by the ultraviolet into ozone, and this acts as an opaque screen, high up, that shuts most of the ultraviolet off from the lower levels of air and from the earth's surface. In early times, therefore, the ultraviolet reached deeply through to cause the formation of the organic molecules mentioned and, continuing to act on them, tended to combine them into still larger, more complex forms. Although the ultraviolet would eventually have disintegrated them also, the heavier molecules tended continually to settle down onto the land and there to become covered up, or, more often, settled into the water.

In these situations the organic molecules were better protected from breakdown, and were subjected, instead, to churnings and

* Investigators at Fordham University are recently (March 1961) reported to have found in one meteorite an abundance of long hydrocarbons, in proportions like those in waxes derived from earthly organisms. They are inclined to the view that these represent the remains of actual living things that evolved elsewhere, rather than preliving organic material.

mixings with one another. Thus the primative ocean, accumulating ever more of them, must gradually have become transformed into what Haldane has called a "soup." And in this soup all sorts of natural experiments in organic chemistry were carried out so as to form ever larger, more complicated molecules and aggregations of molecules. Thus our earth must have acquired all the chemical preconditions for the origin of life.

The Origin of Life *

What the essence of this great event consisted in is a matter that has had much light thrown on it in the last few years. But first let us review what had previously been known along these lines. At the heart of all living things on our planet is the material of heredity. This exists in the form of the microscopic threadlike bodies called chromosomes that are present in every cell. It is these chromosomes that are passed down from each generation to the next. They carry, in their thousands of differentiated parts or genes, which are arranged in them in a line like beads in a chain, the complicated specifications that control, through their chemical reactions, the development of each new body, or offspring. These chromosomes reproduce their like before any cell reproduces and thus enable each daughter cell to grow by fashioning its other materials, or protoplasm, in the accepted pattern according to the given specifications.

And it is also these chromosomes, undergoing occasional sudden changes in their inner composition, called mutations, in consequence of accidental ultramicroscopic encounters that give rise to new varieties, which in turn transmit their new characteristics to their descendants. If these varieties happen, by a rare accident, to be well fitted to survive and to multiply, they can serve as a stepping stone in the long step-by-step evolution of ever higher, more complicated forms of life.

Only in the last five or ten years has it become possible to represent these facts in the formulas of chemistry. It has become

* For a further discussion of this subject, with references, see my article in *Perspectives in Biology and Medicine,* Autumn 1961.—E.A.C.

clear that the inner material of the chromosome, which specifies what the rest of the cell shall be, which reproduces itself, and which mutates to make evolution possible, consists of nothing more nor less than a chain-molecule of nucleic acid. The individual links, called nucleotides, are of only four types in any given case, and it is their exact arrangement in line, single file, that determines the kind of chemical activity, or specification, each group of them has. A nucleotide itself consists primarily of one of the rings or double-rings containing nitrogen and carbon that we have already mentioned, called pyrimidines and purines, of any of their four types, attached to a simple sugar, and this again to a phosphoric acid group.

The way they reproduce, as first worked out by James D. Watson and Francis H. C. Crick in 1953, is by means of each nucleotide fastening down next to itself an appropriate nucleotide of just the right one of the three other types from among all four types that were floating free in the medium about it. The chosen one may be considered as its partner or complement, for each of the two kinds of purines or double rings has an affinity for a particular one of the two kinds of pyrimidines, or single rings. By becoming thus attached to the original chain of nucleotides a whole new row of them, exactly complementary to the first one, becomes aggregated, and these become hooked together into a chain. Later the new and old chains become separated, and each repeats the process of forming a new chain complementary to itself. Now, the complement to the complement is of course identical in nature with the original chain. In this way, through two steps, an exact replication of the original chain has been brought about. This, then, is the crucial process in reproduction.

It has also become evident that what a mutation usually consists of is a substitution of a different one of the four nucleotides in a given position in the place of the nucleotide already there. This could occur either as a mistake in their selection of complements during their reproduction, or by an accidental alteration or "swapping" of material at some point in an already formed chain. Naturally the altered arrangement tends to perpetuate itself through the further acts of chromosome reproduction, and so a new variety may come to make its bid for power.

That even a small chain of a few nucleotides, artificially put together in the laboratory, is capable in a test tube, under suitable conditions (as yet requiring just one enzyme), of gathering to itself free-floating nucleotides from the medium about it, and thus of reproducing itself to an unlimited extent, was recently demonstrated by Arthur Kornberg and by Severo Ochoa. This was the feat which in 1959 won them a Nobel award. But it is an event that must also have occurred naturally in earth's primeval soup.

This step represents the critical one in the origination of life, since from that point onward natural selection of the Darwinian types would in a state of nature inevitably take over. It would do so by selecting those chains in which the mutations in number and arrangement of nucleotides had happened to give them greater stability, together with such chemical influences on other substances about them as would enhance their own powers of multiplication. Thus, gradually, the nucleotide chains came to have ever more profound effects on the molecules of amino acids and other surrounding materials, effects that remodeled, assembled, and integrated them into proteins and other accessory substances helpful to the chains themselves in their competition with one another for survival and reproduction. Thus, step by small step, that enormously complicated system which we call protoplasm was gradually fashioned, as helpful mutations continued to accumulate in the evolving nucleotide-chains or chromosomes.

Despite the inevitability of the process whereby the chromosome primordia, once set going, would gradually come to organize that workshop of theirs that we call protoplasm, whereby they carry out ever more marvelous operations in their own behalf, nevertheless the exact series of steps taken in this evolution are by no means inevitable. For just what mutations happen and become established will depend on many accidental circumstances, and there are surely many different possible pathways that the course of organization could take. Therefore, while we should recognize that life, in the form of material similar in its properties to chromosomes, could arise on any moderately warm planet whose materials were like those on our primitive earth, nevertheless we should also recognize that the further progress of that life into something akin to protoplasmic form, and then beyond that into still higher manifestations, must have had a multitude of different directions open to it. Thus, even though life elsewhere may be based in a core that is much like a chain of nucleotides, and though this core may have established chains of amino acids that we call proteins (some of enzymatic action) to serve as its second-in-command over the operations of the now living matter, nevertheless it is to be expected that from there on, if not earlier, the biochemistry would be radically different, with alien substances in the place of our vitamins, sterols, and so on.

The Tell-Tale Cues on Mars

Corroboration of the conclusion that life arises on any planet on which even half-way suitable conditions exist is provided by observations of the planet Mars by astronomers. There it is far below freezing over most of the surface most of the time, free oxygen is very nearly or completely lacking, and water or water vapor exists only in minute amounts. Nevertheless, the drought is not complete, and the temperature becomes temperate over part of the surface towards the middle of each day. Thus a little life of a kind based, like ours, on nucleotides and proteins in a water solution would seem to be possible.

As has long been known, telescopic observation shows fairly extensive dark grayish areas, although most of the surface is of a reddish-yellow hue suggestive of sandy deserts. Moreover, the dark areas, unlike the reddish ones, change in shade with the seasons in a manner very suggestive of the grayish-green vegetation that exists at high latitudes and high altitudes on our earth. For when winter approaches on Mars these darker areas turn brownish, and as spring advances they become gray or grayish-green again. It is especially noteworthy that in summer dust or sand seems occasionally to be blown over parts of the dark areas, obscuring them, yet before many days elapse the overlying layer seems to fade away as though it had somehow become swept off or grown over.

At the observatory of the Pic du Midi, in the Pyrenees of France, observations of the polarization of the light reflected from the dark areas, at various angles in relation to the angle of the incident light, have given evidence that the surface is peppered with a host of tiny objects. It can be reckoned that in winter these are somewhat smaller, on the average, than a tenth of a millimeter, so that if we were on the spot they would be only just visible to our naked eye. But as the season advances towards summer the polarization changes in such a way as to lead to the conclusion that these objects have grown to about ten times their winter size. And so it goes, back and forth, as the seasons succeed one another and as the coloration changes concurrently.

The most recent findings are those that have been made by detailed examination of the infrared light reflected from the two kinds of areas of Mars. In these examinations the light, after having been received through a telescope, is passed through a spectroscope in the usual way, in order to be spread apart into its different wave-lengths. It is then found that there are dark bands, indicating the absence of light of given wave-lengths, in particular positions along this spectrum. The missing light-waves had of course been blocked by the material on the planet's surface. Now the position of these bands corresponds exactly with those of the dark bands obtained when infrared light is reflected from certain known compounds here on earth and examined in the same way. In the case of Mars, these absorption bands are found only in the reflections from the dark areas. As for the compounds on earth that give the same bands, they are all organic compounds, of a type which on earth today are manufactured only by living things, and the structure in these compounds responsible for these bands is a certain type of chemical bond between carbon and hydrogen. This is as deep as we have yet been able to see into the nature of our sister planet's mysterious dark areas.*

It is true that no one of these lines of evidence is conclusive by itself. Taken together, however, they make up a very weighty body of testimony for the presence of living things on Mars. Now con-

* According to letters from N. B. Colthup and W. M. Sinton in *Science*, 25 August 1961, two bands point especially to acetaldehyde, which on earth is a product of anaerobic catabolism, while another band seems to indicate carbohydrates or proteins.

sider that this is the only planet in our solar system, other than our earth, on which the conditions of temperature and composition would allow living things composed of carbon compounds to exist actively. Then the fact that these consistent signs of life have been obtained from this very planet, despite the *relative* unfavorability of its conditions, is seen to constitute a powerful argument that life will arise and evolve anywhere in the universe that a setting exists which is only half-way suitable for it to struggle along in.*

No doubt the life on Mars is far more meager and more primitive than that on earth. Not only has the room in which it could live been far smaller there, but the lower temperature, lesser illumination, and dearth of water and atmosphere must limit the rate of chemical turnover to a small fraction of ours, and along with this must correspondingly limit the rate of evolution. True, there must once have been more water on Mars, before most of it had evaporated off, and life must then have been more flourishing, though incomparably less so, at its best, than on our earth. But our earth in turn may be poor in life compared with what exists on some of the even more favored planets of other stars.

The Possibility of Life Based on Other Materials

In any place in which living things have evolved, the primitive living matter must, like the chromosomes we know, have the faculty of replicating, by arranging particles taken from the medium about it into a pattern like its own, and the faculty of having

* However, Edward Anders has recently argued in *Science,* 14 April 1961, that oftener than once in a million years the earth is hit by asteroids or meteorites so large that their impact could throw off fragments of the earth's surface with escape velocity. Some of these, harboring microorganisms, could seed other planets in our solar system, and possibly even planets of other stars. Reciprocally, our own microorganisms might originally have been derived from elsewhere. Thus, actual biochemical comparisons will be needed for throwing light on whether life forms on earth and Mars, or on any two accessible planets, are of common or independent origin.

occasional mutations in its pattern, which are then subject to replication themselves. For, as we have seen, these procedures inevitably result in that accumulation of helpful mutations whereby more highly organized beings gradually take shape. However, we may well ask, could there be some other kind of potent chemical materials than nucleotide-chains that like the latter are able, under some circumstances, to engage in replication, and to undergo mutations some of which are helpful to their replication, and that consequently have the tendency to evolve to ever higher forms? In other words, must all life everywhere be based on nucleotides that construct proteins and work in a water medium?

It has been imagined, for example, that under much colder conditions, where ammonia is liquid but water solid, there might be complicated replicating molecules other than nucleotide-chains, immersed in the ammonia, and serving as the basis of living things quite different in their building blocks from those we know. Again, at temperatures so high that water is vaporized, might not some type of compound based in part on silicon instead of carbon become organized in a still different medium, liquid at that temperature, so as to form still another type of life?

Although such speculations cannot be categorically dismissed, there is no factual basis for them, and no combinations of compounds and mediums have been conceived that could work as imagined. It is true that the relevant chemistry for forming a judgment is as yet very undeveloped, and that we have only recently understood the bare outlines of how our own nucleotides operate to make our own type of life possible. Nevertheless, even on our earth very diverse physical and chemical conditions occur in different situations, and we have nowhere found a suggestion that any other kind of life exists or has existed anywhere on earth, other than that based in nucleotide-chains. Obviously it is the highly special structure of these nucleotides that gives them the faculties on which the evolution of our life has been based. We may therefore "keep our fingers crossed" about the possibility of life of wholly different constitution than that on our earth having arisen elsewhere, while at the same time we should admit that in the light of present knowledge this hypothesis seems highly unlikely.

As for the nature of life that exists under a range of conditions not too different from those on earth, we should within the lifetimes of most of the readers of this article have some solid facts to help us judge whether or not such life is likely to be founded on material different from that of nucleotide-chains. For surely, if civilization manages to avoid destroying itself by war, we will long before the turn of the next century have succeeded in reaching Mars and in probing into the nature of its organisms. Thereupon will begin the most exciting story in the exploration of life that has ever happened to man, except of course for the story that is going on right now in those laboratories of ours where biochemists and geneticists are disentangling the warp and woof of which our own earthly life is composed.

Even if the living matter of Mars and of all other worlds on which life occurs has been restricted in its origin to chains of nucleotides, or to something much like them, working in water, there would still be very different pathways to its further evolution. By analogy with earth's life we would expect that the next step would be the action of the nucleotides in picking out from the medium molecules of amino acids, which are the building blocks of protein, and the stringing of them together to make protein. But we cannot be sure. And even if it were the case, it would not follow that all these amino acids were just the same in their construction as the twenty different kinds that our nucleotide-chains make, for there are many more types possible. One difference might be that on Mars some or all of the amino acids in the proteins were mirror-images of ours, twisting light to the right instead of, like ours, all to the left. This is but one little example.

Moreover, as has been mentioned previously, the secondary, more specialized materials, including vitamins and other prosthetic groups, as they are called, hormones, toxins, etc., and even the common materials corresponding or analogous to carbohydrates, fats, and other lipids, are likely to be much more different from those of earthly organisms than those of earthly organisms are from one another. For those in earthly organisms are on the whole in surprising agreement, as if the common ancestor of all species on earth had already acquired most of our own biochemical composition before the tree of evolution put forth the branches that we find on earth today. Thus, while the substances of practically all other organisms, animal, plant, or microbe, found on earth can be digested and used by us, at least after cooking, being interconvertible one into the other, this is not to be expected of life forms evolved after an independent origination of life on any other planet. We would not be able to use them as food nor could they use us. In fact, we would likely be mutually poisonous to one another.

Some Expected Points of Agreement

However that may be, life on any other planet having our range of conditions would eventually have come to derive nearly all of its energy by the absorption of the sunlight by means of pigments. Most likely these pigments would be porphyrins, like those in chlorophyll of earth's plants, although it seems unlikely that the complicated structure of this chlorophyll would be exactly duplicated somewhere else. Once this means of capturing energy and therewith synthesizing the compounds needed had been achieved, there would inevitably be a tendency for organisms to split, as they did on earth, into the synthesizers or plant-like types and the predatory or animal-like ones. The development of plant-animals combining both modes of life that one sometimes reads about in science fiction is very improbable. For an enormous surface is needed to gather enough light to maintain a fair-sized animal, but such an organism would not be compact enough to have the movements necessary for an animal. Thus for plants above microscopic size there must be an emphasis on reaching out and absorbing the minerals, water, carbon dioxide, and sunlight needed. In the case of land plants this means that there must be roots for the minerals and water, leaves for the carbon dioxide and sunlight, and a supporting structure between to hold them together and conduct materials between them. However, within these specifications there can be an enormous latitude of types, especially with regard to means of disseminating the products of reproduction.

In some situations, size is an advantage to a plant or animal.

However, it raises difficulties. For example, simple diffusion of materials is no longer effective enough in transport of materials over distances larger than microscopic ones. For larger masses, compartmentalization into more or less specialized units, or cells, with spaces between or within them for transport, becomes advantageous. We would therefore expect this multicellular condition to develop on any world having highly organized life. At first the multiple cells would tend to be arranged in thin layers in order to allow sufficiently ready in-go and out-go of materials from their surfaces. Further complication would tend to lead to foldings of these layers and migrations from them, as we find in the development of our own higher organisms from the egg to the adult stage.

In the evolution of animals, unlike that of plants, the emphasis tends to be on effective motor response to allow them to get food and to keep them from being themselves used as food. Thus it is advantageous for the more advanced animals to develop improved sensory and coordinating systems and to have a readily maneuverable and powerful and therefore fairly compact body. The resulting tendency to bulkiness calls for increasing servicing of the inner parts, that is, for the development of systems, much more complicated than those of plants, for taking in and processing the food and fluids, for converting and distributing them, and for collecting and eliminating the wastes.

Multiple Solutions for Higher Stages

But although higher organisms on other planets would almost inevitably have undergone all these developments in some form, they may be expected to have followed radically different courses in regard to many of the features. Illustrations of such dissimilarities among fairly advanced animals of widely different types on earth, such as starfish, beetle, octopus, and fish, are familiar to everyone who has studied a little zoology. How much greater, then, might such differences be between the forms of earth and those of another planet. These differences, affecting their whole internal economy, including the biochemistry within their cells, would also be expressed in their gross anatomy and in their outer form.

Just as in the evolution of the biochemical system, so too in that of organs, tissues, and the general ensemble of any organism of a moderately high or high stage of advancement, innumerable steps of progression have been mounted in succession. Among these steps there have often been some that later proved to have been deflections or even descents from the general course. Features have also been included that met some more immediate need but after a while, as when the organism's way of life or other features had changed, turned out to be useful in a very different way and thereupon became subjected to further evolution in that other direction. For natural selection cannot see ahead to later possibilities. It is opportunistic and therefore subject in some situations to long periods of marking time and to occasional lucky turnings of a corner that are followed by a burst of progress along one or more new lines.

Just what step will be taken at a particular point is sometimes a matter of accident: of what mutation manages to take hold, and then what combination of mutations, until some novel structure or manner of functioning is thereby brought into being that acts as a key to open up an important new way of living. This in turn can usher in a whole new series of developments, centered about the given key innovation. A frequently cited example of this kind of thing is the appearance of a structure that can begin to serve for transportation through the air and so initiates the evolution of wings and, secondarily, of all sorts of other changes in adaptation to the life of a flier.

Another unusual key change that occurred in the remote ancestors of the starfish group (echinoderms), was the formation of tubular projections (tube feet), arising from five or more arms that radiated out like spokes on a wheel. The arms were used in getting food, and the tube feet, which passed the food along to the mouth in the middle, were worked by means of changing water pressure inside them. This pressure was regulated by the contraction or expansion of syringe-like bulbs inside the body that were connected with the tube feet. Only on this one occasion in our animal kingdom of earth did this combination of structures arise. However, having once arisen, it was eventually taken advantage of both for grasping and pulling (when suckers were developed at the ends of the tube feet), and for a distinctive kind of locomotion, combined with aggression. It made possible, for example, the overcoming of such slow-moving prey as clams. In time, the form and all the other bodily systems of these animals became radically reorganized so as to take better advantage of these unique tube feet, and so a whole phylum, or great group of species, took shape, that are modeled along lines quite different from those of any other earthly creature.

Another key structure has been the jointed external shell, continued on to appendages also having joints, that arose in the group ancestral to crustacea and their relatives. To this system of jointed armor very much else in the advanced organization of these active aggressors is adjusted.

Contrasting with this, the mollusks, which in the distant past had come from the same stem as the crustacea, did not have their shell extend to their appendages. Instead, in the more typical mollusks the shell early became united into one or two large protective covers, attached to a fold of skin or "mantle," more or less separated from the body, that enabled the shell to envelope much or all of the body. Effectively defended in this way, these animals could largely dispense with high mobility if they concentrated on the scouring of the water or surfaces about them for microscopic prey or plant material.

However, one branch of the mollusks developed an important new key structure, in the shape of uncovered protrusions or tentacles that projected in front. These tentacles gave their possessors (cephalopods) such great potentialities for active and mobile aggression, as well as for defense, that the shell, increasingly a liability for these creatures, dwindled in the higher types. At the same time, such an impetus was thereby given to the selection of ever better motor, sensory, and nervous systems, and to the efficient servicing of the comparatively massive body, that the resultant group of squids and octopuses may be considered to be just as highly developed, after their fashion, as the group of back-boned animals (vertebrates) to which we are proud to belong. However, since their organs have evolved so independently of ours, their anatomy, in spite of showing many striking analogies to ours, is built on a fundamentally different pattern.

Our own vertebrate pattern can be properly understood only by having in mind that it started out as a torpedo- or eel-shape supported by a somewhat flexible inner rod; that it was soon provided all over with a protective covering of teeth, which in the jawed mouth, by becoming enlarged, furnished a very effective instrument of aggression; and that for purpose of stabilization in swimming, paired fins were developed which could later become modified into limbs. Here too, then, was a basis for the further improvement of parts in the interests of better maneuverability, better powers of offense and defense, and reactions ever more effectively adjusted to the changing events in the environment. In this group also, therefore, there was a special premium put on developments that improved the sensory and the nervous systems, and in this way we ourselves at last came into being.

The Bizarreness of the Right and Proper

Many key developments like those we have been considering seem to have arisen in a successful way only once or twice in all the history of life on earth, and their beginnings must therefore have involved a very accidental combination of circumstances. This being the case, there must have been very many more key developments possible, of just as great potential significance, which never did occur on our earth, at least not in a successful manner. On any other planet that supported life one would therefore expect to find some other developments, together with all the accessory features that they had given occasion for.

Thus, even without taking into consideration the great differences in the biochemical basis and the lower evolutionary stages between life somewhere else and that on earth, the higher developments there would be expected to be at least as different from ours in their general pattern and workings as the ordinary dog, the tarantula, and the chambered nautilus of our world are different from one another. Certainly, then, we could not classify the organisms there within any of the earth's grand categories or phyla, such as vertebrates, mollusks, or flowering plants, much less within any of the narrower divisions, such as the classes, orders, families, or species known on earth. Viewed in this light, we see how utterly foolish it would be for us to expect to find human beings to have evolved on any other world.

It is most naive of human beings to assume that the human type, or any other type familiar to them on this little planet, is the right and proper type, and that nature has been subjected to some sort of irresistible compulsion to produce it. Looked at in a larger light, we humans and all our companion species of here-existing animals and plants are most bizarre, as any visitor from elsewhere would readily testify to. That is, although our organizations are full of intricacies that work wondrously in relation to one another, and although the over-all structure of some parts, such as the eye, is the only workable arrangement that could have been developed from the basis with which it began, nevertheless many other features of the pattern represent very gratuitous arrangements or makeshifts that we try to make the best of.

Take, for instance, the way we and other land vertebrates breathe air. It is very inconvenient to have it come through the mouth or, coming through the nose, that it should have to share a part of the same passageway as must serve also for food. Thus, the land snail, whose lung has a passageway and opening quite distinct from its food canal, is much better off in this respect. So too is the grasshopper or other insect, which breathes through portholes placed near the organs to be aerated, or which as in the case of aquatic larvae breathes through openings located at the hind end, that can project out while the mouth continues to feed comfortably down below.

Our own inconvenient arrangement harks back to the time when as fish both our oxygen and our food were in the water that came via our mouth, and we gulped air for our swim bladder. In non-vertebrate water-breathers, on the other hand, the gills have not been in slits of the food canal and have no connection with it except in the case of the sea-cucumbers that breathe through their rectum. In most other gilled animals the gills, being tender, lie tucked away under the abdomen or protected in a fold or under a mantle.

Let us proceed a little further in considering the strange combination of features that our own mouth represents. Besides acting to take in both food and oxygen, it serves the majority of the vertebrates as their most powerful weapon and at the same time as their deftest organ of manipulation activities usually assigned by other progressive phyla to specially constructed appendages. In addition, in most land vertebrates the mouth has the peculiar function of emitting sounds, while other animals that make noises do so by quite different means, as by fiddling, buzzing, or strumming.

Finally, in human beings and their relatives the mouth is also used for the important purpose of expressing their feelings, as through scowls, sneers, smiles, and so on. Although these modes of expression seem natural to us, being inherited in us and even present in other mammals, as Darwin pointed out in his fascinating book, *The Expression of the Emotions in Man and Animals,* nevertheless the meanings they and our other bodily signals convey to us would not be apprehended by any alien creature who had not had an opportunity to learn them.

In short, then, the alien would find it most remarkable that we had an organ combining the requirements of breathing, ingesting, tasting, chewing, biting, and on occasion fighting, helping to thread needles, yelling, whistling, lecturing, and grimacing. He might well have separate organs for all these purposes, located in diverse parts of his body, and he would consider as awkward and primitive our imperfect separation of these functions. Thus, for the science fictionist to picture the alien with an absurdly all-purpose mouth like ours, as he usually does, is a prime example of the prevalence among our supposed intelligentsia of utter unsophistication in regard to basic biological matters.

Our example of the mouth is by no means the most striking of its kind that might be given. The whole concept of the head as we think of it is pretty much a man-centered or rather, a vertebrate-centered one, as one realizes when examining a spider, octopus, or crab, in which the head and much of the trunk are merged. Moreover, neither the organs of hearing nor of smell need be near the brain or the mouth—as witness the grasshopper, which has its ears on its abdomen, and the fly, which smells and tastes with the tips of its feet. Again, consider the scallop, a headless but alert mollusk that can swim actively and whose many eyes, structured much like ours, are arranged in a long row on the mantle, Just beyond the edge of each shell. So we could go on and on, citing the seemingly *outré* of this very planet earth as evidence of our own singularity.

There are other important object lessons to be derived from a comparison of sense organs, and more especially of eyes. Advanced eyes built on principles basically like our own, complete with lid, cornea, adjustable lens, iris diaphragm, retina, absorbing layer, etc., have been evolved independently in several very different lines of descent, among them the octopus-squid group and, though less perfectly, the group of scallops just mentioned. They proceeded from similar beginnings but took a different evolutionary course, whereby they reached, however, essentially the same final form. Nevertheless, eyes using entirely different principles of image formation have also been evolved, such as the convexly radiating set of tubes that form the insect's compound eye, the pinhole type of the nautilus, and the scanning eye of the snail. But we know at least one other principle that would be effective, the reflector type that is embodied in some of our telescopes, which, as it happens, has never yet been hit upon by earthly organisms though they do use accurately curved reflectors for some other purposes.

As for the kind of radiation to which known eyes are sensitive, we may note that many animals, for instance bees and some butterflies, are sensitive to a range of colors which, by comparison with our range, is displaced toward the shorter wave lengths, so that they fail to see the reds but do see well into the ultraviolet that we are blind to. The snail's eye is very sensitive to X rays and other ionizing radiation. The rattlesnake and its relatives have delicate directional infrared detectors in the two special pits on their head, by which they can sense the presence and position of mammals and birds. The king crab is one among various creatures that can distinguish the direction of polarization of light, somewhat as we can its color. Water currents and vibrations of different types, unlike or beyond the sound waves we hear, are also sensed by special organs in diverse organisms. And there is no doubt that there are in earthly creatures senses, and elaborations of senses, that we are as yet unaware of.

I draw the line, however, at the idea of telepathy for any earthly creatures as at present built. It would take me far too long to explain my reasons for this view, except to say that it implies phenomena quite contradictory to all we know about the ways of working of the nervous systems of man or earthly animals. Nevertheless, there might well be natural means of communication between creatures that had developed elsewhere, using physical channels of transmission and reception to which we are insensitive. We might at first regard these beings as telepathic. Similarly, a group of people all of whom were deaf might infer that we communicate telepathically with one another when we speak.

In the matter of communication, we are apt to take it for granted that the natural method is by sound, although we do use sight also. Ants convey messages through touch by motions of their antennae. Bees use something akin to pantomime. For all we know, squids may be signalling very elaborately when they control the colored patterns that shimmer over their surfaces in such varied forms, and it is even possible that they can do this at night and in the dark depths of ocean when they cause their skin to light up with moving fluorescent designs. All this should make us more broadminded as well as cautious when we try to imagine the ways in which a fairly intelligent creature of another planet would behave.

A Few Rules of Inference

There are, of course, some principles that can help us in inferring what types of life might have developed on a planet of known temperature, mass, and composition. The smaller and less favorable the area on it for the flourishing of life, and the shorter the time since life could begin on it, the less advanced and the less diversified that life will necessarily be. Again, the greater the planet's mass, and therefore the force of its gravity, the shorter and stockier its land animals will tend to be, to bear their own weight, while conversely lesser gravity would tend to lead to slimness. Greater gravity would however tend to go with an atmosphere disproportionately dense, even in relation to the weight of objects, and so it would be favorable for the development of flying, and perhaps even for balloon-like lifting organs, secreting hydrogen or methane. However, we must remember that the force of gravity is relatively unimportant for aquatic forms, since the water tends to support them.

Most of earth's highest developments among animals, and more especially among plants, took place in the more varied and variable environments to be found on land. Yet life must have its origin in water. Therefore, other things being equal, the higher developments of life will be found on those planets that have considerable areas of both water and land.

Such planets might be very exceptional, however, if the hypothesis, now somewhat out of favor, should prove to be correct after all, according to which our great ocean basins trace their origin to the tearing of the moon out of the earth's crust in the Pacific area by means of the sun's tidal action at an early time. For it was the same process as dug the ocean basins that conversely must have left the continents high and dry. However, only certain very exact quantitative relations between the earth's and sun's mass and the period of the earth's rotation on its axis, could have led to a cataclysm like that here conceived. Sir George Darwin believed his reckoning showed that the necessary relations did hold in this instance. But, if so, an event of this kind could have happened very seldom elsewhere. Hence, if this is really what it usually takes to raise continents, very few planets indeed must be provided with them. The moons attending other planets in our solar systems were almost certainly not formed in this way. On this theory, then, life on the great majority of planets would be confined to aquatic types, plus any with drastic innovations which had enabled them (like flying fish) to soar out of the water or (like the argonaut or masses of seaweeds) to make floats for themselves or for organisms associated with them.

What of Intelligent Life Elsewhere?

We may be sure that in the competitive struggle of higher animals for existence, on any world, the effectiveness of their sensory and coordinating systems will play an important role in determining their success. For the more effective these systems are, the more likely are the animals to carry out movements that enable them to take better advantage of opportunities presented by their environment, and to guard themselves better against its dangers. Thus, through the survival of mutations helpful in these respects, their sensory and nervous systems will tend to accumulate a succession of improvements, including reflexes and instincts that result in ever more serviceable reactions.

However, as we know from the evolution of behavior in our own world, much additional advantage may be gained if the nervous system is so constituted as to allow the modification of these reactions in conformity with the individual's own experience. This process, the neurological basis for which is not yet understood (although it is said that the results of it may to some extent be imitated in certain artificial computers), is on its lower levels termed "conditioning" and, as it becomes more highly developed, "association," "conscious learning," and, finally, "intelligence."

It is hard to judge how much of an innovation is required for establishing the *rudiment* of the mechanism that allows conditioning. For we do not know how far down in the scale of our animal life it is present and so do not know if it arose just once or had a multiple origin. Yet where it is very rudimentary, as in some worms, it seems to be of so little use that it would hardly have become established in evolution unless it had at that time involved a fairly simple development. At any rate, we do know that as evolution continued several different lines of animals that we may call progressive all carried the development of conditioning to a far higher degree. These included more especially the lines with jointed appendages (notably crustacea, insects, and the spider group), the active tentacled mollusks (squids and octopuses), and of course our own group of vertebrates. In view of these multiple parallel developments, all of them such notable ones, it seems very likely that in the evolution of higher, more active animals on practically any world, association, learning, and intelligence would at last make their appearance.

Even such higher levels of understanding as are evidenced by the carrying out of constructive operations nicely adapted to individualized situations, the use of inanimate materials as tools, and the exercise of foresight in making plans are, on this earth, by no means confined to man or even to mammals or the vertebrates. To refute the idea that they are, we need only call attention to the engineering of the beavers, the way some birds build their nests, and some spiders their webs, to the strategies used by wolves, foxes, and raccoons in hunting and in eluding pursuit, to the way a wasp on occasion selects a pebble and then pounds down the earth on its nest with it, or to the practice of the octopus in dropping a stone between the shells of an open clam so that it can then safely insert its tentacle and extract the soft body. There is much more than mere instinct in all these activities.

However, it is only in our line of descent, of all those on earth, that intelligence has advanced far enough to allow elaborate communication to take place, and that at the same time social feelings have become so well developed as to result in the ever more effective transfer of understanding and skills by example and precept, from individuals to their companions, and from elders to youngsters. This is a process which, in the large, we call "cultural evolution," for it goes on and on. In other words, biological evolution has to proceed very far before the native intelligence, combined with sociality, reaches such a level as to permit the social advances which engender civilization and eventually science. Yet I am convinced that, if man should disappear, a creature something like one of those I have named would eventually succeed in attaining that level on its own. But so long as man is here, he holds that door closed to it.

On another planet having conditions as favorable for animal life as ours, there is every reason to expect that in a comparable period evolution would have resulted in as high intelligence as in our own more intelligent animals, and that in time one of these would have stepped over the threshold into the realm of culture, civilization, and science. As we have seen, such a creature could not be anything like man or even a vertebrate in his bodily construction or physiology. Moreover, on different worlds he would surely be most diverse in regard to form, general behavior, and methods of natural communication (although he would surely have some such methods). To picture him as probably bipedal reveals a naive egoism, although he would of course have manipulable appendages that were freed from the duties of locomotion. As for his intelligence, however, despite its independent origin, it would certainly be capable of achieving much mutual understanding with our own, since both had been evolved to deal usefully with a world in which the same physicochemical and general biological principles operate.

Would he in most cases be ahead of us or behind us in civilization, knowledge, and techniques? That is a question which a little consideration will show to be easier to answer than it would at first sight seem to be.

Let us first recall that known fossils show our earth to have had cellular, photoplasmic life more than two and one-half billion years ago. However, men with a culture higher than that of food gathering seem to have existed for not more than 12,000 years. This, then, is only one two-hundred-thousandth part, or less, of the whole time our biological evolution has consumed. But astronomical evidence indicates that our sun and earth are likely to continue with approximately their present temperature and other natural physicochemical conditions for a period far longer than two and one-half billion years into the future. Thus, if man uses his great gifts wisely he is likely to be able to continue his march at least as long as life has already existed on earth. During all that time he would have the opportunity to continue advancing in knowledge, in techniques, and in regard to his inner being and his social relations, at a self-controlled pace. Undoubtedly he would also extend this progress to his own genetic constitution, as a result of his own free choice to do so. This genetic advance would in its turn allow him to reach still higher levels of civilization and of control of the world about him and himself.

Now if, as some astronomers conclude, many of the stars and planets in our galaxy are much younger, and many others much older than our sun, with our sun not far from the average in its age, it is easy to see that worlds having life, if similarly distributed in age, would include about as many older as younger than ours. If then their *average* age was some five billion years, as ours is, and their range of ages such that the average difference between two taken at random was one billion years, we can see that as often as not another life-bearing world would be a billion years behind us or a billion years ahead of us. A billion years behind us would probably put us back at about the level of jelly fish, but a billion years ahead—what would *that* be?

In the light of these considerations, the finding of a planet whose highest life-form corresponded with something between our beginning civilization of 12,000 years ago and one of 12,000 years more advanced than ours would be like having the good fortune to draw the one lucky number out of 80,000. For this range of 24,000 years is only an eighty-thousandth part of the 2 billion years range—a billion in either direction—that we are likely to be dealing with.

In other words, in only one out of 80,000 planets having life would we be likely to find it at a stage having a civilization at all comparable to ours (though personally I would not think of a civilization a mere thousand years ahead of us as "comparable"). In half of the rest life would not yet have reached the stage of civilization, usually not that of any culture at all. But in the other half it would be so far ahead of us that we would stand completely abashed before it. What that kind of contact would mean for us is a matter that might be speculated about at great length but with little satisfaction.

Why Have We Found No Sign of Them?

There is, however, one very jarring note in this composition. If there are really so many millions of worlds in our galaxy whose inhabitants are so far ahead of us, why have none of them already settled here or at least given us any evidence that they visited here? All our life has obviously been derived from but one tree. But everywhere the law of life is to expand and to settle wherever it can. And though we are at present preparing only for interplanetary flight within our solar system we may be sure before many generations have elapsed we shall be ready even to tackle the far vaster distances between the stars. If *They* can do all this so much better, why have they not already done so?

Before closing let us briefly go over a few of the more obvious possibilities. One is that the physicochemical organization that makes possible conditioning of a type that can later be elaborated into intelligence represents after all an extremely improbable combination of mutations. On this view, it happened only once on earth, and the intelligence of all the diverse types of higher animals mentioned really traces back to that unique beginning. So exceptional, on this view, would this event have been that it took place on only this one planet of ours or at most on only a very few of the myriads harboring life.

Then there are the cynics, the standpatters, and the oldsters who would maintain that the distances between the stars are really too great to be conquered even by the efforts of the most intrepid explorers willing to make the sacrifice of spending generations on the way. We would ourselves admit that almost surely the dream of exceeding the speed of light is an idle fantasy, like that of perpetual motion. And the time necessary for traveling at a lesser speed poses for these objectors an insuperable difficulty. Some might also maintain that the cosmic rays are too destructive (although if one had enough energy one could send a body so massive as to provide sufficient shielding). And as for other dangers, they say, we just have not discovered them yet.

Another line of thought would have it that an intelligent being after reaching a somewhat advanced stage inevitably destroys himself, if not by mass warfare or genetic neglect, then perhaps by sinking to a decadent complacency in which he philosophizes himself out of the motivation to continue his expansion or even his life. Or he might construct such advanced robots that they at last took over and yet lacked some requisites for the unlimited continuance and expansion that is characteristic of life itself.

All these interpretations are essentially pessimistic for they would limit us as well as other life forms. There is on the other hand the optimistic possibility (however improbable it may be) that we happen to be ahead of anybody else, or at least of anyone near enough in our neighborhood to have reached us as yet. In support of this idea it has been suggested that our own sun, although far from being one of the oldest stars, is one of the oldest in that third or fourth generation which alone would supply enough of the heavier elements that are needed to make possible such active life as ours. In that case we would have the edge on the others in evolution, and we will get to them before they can get to us, but they will have nothing to teach us except what we can learn by our own efforts at studying them.

Finally, let us recall another circumstance that might have given us the edge on them. If most planets that could support abundant life are rather evenly contoured, as Mars is, and are given the major inequalities of level necessary to differentiate continents from oceans only by such a cataclysm as George Darwin pictured in the supposed tearing off of our moon, then nearly all these other planets are almost completely covered by water. If that had been the case on our earth, then, as we have noted previously, most of our highest types of life, being land types, would not yet have had the opportunity to evolve. By the same reasoning the other planets that lacked continents would also have been retarded in their evolution. Whether this retardation would be sufficient to have kept even those one or two billion years older than we from reaching our present level is another matter that we might at present speculate upon fruitlessly.

Whatever the answer may be, I believe that we shall some day—and before a thousand years are up—find out. In the meantime, we shall have plenty to do. Let us make sure, however, that we do not use these great new powers and new insights in mutual destruction. Let us also make sure that we preserve that spirit of free inquiry, free expression, free criticism, and humane conscience which should make it possible for us to challenge space without dehumanizing ourselves in regard to our most essential values.

The Meaning of Freedom

THE word "freedom," like the sleigh of Santa Claus, is heavily loaded with much that is dear to our hearts. It is fraught with associations that bring feelings of escape, exhilaration, adventure, power, pride, and fulfillment. Consequently we must remain on guard, if we are not to be bewitched into supporting any policy to which this magic word has become attached. We must beware of approving betrayals of freedom that bear its mask, such as so-called "freedom to work" laws, "freedom to teach religion in public schools," or "free enterprise" in the sense of the right to form combinations in restraint of trade. In short, we must examine freedom dispassionately, groom and guard her vigilantly, if we would win the right to embrace her passionately.

Fatalism

One of the oldest misconceptions concerning freedom is that it is opposed to the operation of cause and effect. Rather than violating natural laws, it constitutes our opportunity of utilizing these laws as we ourselves desire—that is, in what we conceive to be our own interests. Both oriental fatalists, believing that men are prevented by cause and effect from influencing their destinies, and advocates of that mystical idea of "free will" which allows men to rise above or circumvent physical principles, are committed to the same false antithesis between law and freedom. They fail to see that if events in general were not lawful we could not foresee what the effects of given actions would be, and hence we could not control the outcome of events. They also fail to see that if our own decision-making processes were not lawful there would be no way that our decisions could be steered into directions conducive to our interests.

In spite of the modern physical principle of indeterminacy, thinkers of the ancient East did arrive at a profound generalization when they maintained that the world of long ago contained, in the arrangements and motions of its parts, the makings out of which today's world lawfully issued. But those groups were wrong who concluded from this that men might as well sit idly by watching calamity overtake them since they are impotent to decide any outcome. For men are themselves an outgrowth and a part of nature, and nature works through men as truly as it does through sticks and stones, or through storms and stars. And an integral component of men is their power of making decisions in accordance with their own natures.

Thus the muddle-headed doctrine of fatalism has blinded itself to the only genuine kind of freedom, lulling its adherents into an apathy in which they fail to avail themselves of existing opportunities and fail to create new ones. This attitude has been an opiate just as surely as have religions that offered the down-trodden their reward in an afterworld. The resulting submissiveness of the people at large has, in the case of both illusions, pleased the powers-that-be by leaving them a relative freedom denied to the majority.

The classic "free-will" advocates, accepting the fatalists' misconception that natural law, if inviolable, would make man helpless, have attacked the idea of its inviolability and taught that somehow the will overthrows the order of nature and cuts the Gordian knots of physical law. Thus the adherents of this doctrine have generated a contempt for science. Then, as science has continued to develop anyway, they have seized at every temporary gap in its fabric and inserted hooks on which to hang their notions of spiritistic interventions that deflect and dominate physical phenomena. Some of these are supposed examples on a higher plane of what the free will of man may accomplish in its smaller way.

Genesis

Among the foremost instances of such fancied operations of supernatural free will have been the acts of genesis. In present-day science, the origin of the visible universe still remains a field for conflicting theories. According to one version it arose ten or more billion years ago through a titanic explosion, that may in turn have been preceded by an equally stupendous contraction or collapse, and these movements may have continued in vast cycles. But according to Hoyle (whose name belies his iconoclasm!) matter is continually emerging unobtrusively in tiny bits throughout space, and there is no beginning or end, or cycle of things as a whole.

Yet, whatever group of astronomers and physicists may prove to have the better case in this argument, virtually all of them now agree that our own solar system originated by the gravitational falling together of diffuse gas and dust some five billion years ago. They agree that in this process our earth and the eight other visible planets and their satellites arose as local condensations while the main body of the solar cloud shrank to form our present sun. Moreover, modern evidence indicates that a considerable proportion (perhaps 10 percent) of the hundred billion stars in our galaxy, as in the billion or more other galaxies, have developed their own families of planets. Our knowledge of the physical properties of matter is already advanced enough to lead to the expectation that scattered matter—of which a good bit has been left dispersed and of which more is still being strewn about by stellar explosions, if not in other ways—would inevitably gather together by these means, so as to result in the stars in our firmament, the ground beneath us, and our air and oceans.

The next great mystery of genesis concerned the origin of life. Most peoples of the past who thought about it agreed that here too direct intervention by a supernatural will was necessary. This will breathed into inert matter a vital sentient spirit, something akin to itself, which in its turn could will. Moreover, life was supposed to have been breathed separately into each of the kinds of animals and plants that are thought of as different species.

Nearly a century ago Darwin suggested, and E. H. Haeckel

urged, that organic compounds had been formed in the primitive waters by natural chemical processes, and that these interacted with one another to form the first beginnings of living matter. Now modern physics and chemistry are drawing the outlines in far better detail of how radiation, electric discharges, and heat, acting in the churning outer envelopes of planets, must have built up an ever increasing horde of complex molecules of the kinds that constitute the building blocks of living things. Some of the most important of these reactions have in fact been imitated in the laboratory.

The most essential of these building-block molecules are known as nucleotides and contain phosphate. In the last fifteen years, research has shown that these nucleotides, of which there are four main kinds in any given case, come to arrange themselves in a chain, and that in this arrangement they constitute the central organizing core of living things. This is the material of the microscopic threads called chromosomes that are the basis of reproduction and heredity. These chains of nucleotides have been found to be capable, even in a test tube, of directing the assemblage of scattered individual nucleotides from the fluid around them into new chains that are for the most part, in consequence of two main steps, perfect duplicates of the old ones; having the same pattern, like letters in a word, of the four kinds of blocks. Thus the chains give rise to other chains like themselves—that is, they reproduce.

However, changes or "mutations" do sometimes take place in the arrangement of nucleotides in a chain. These substitutions of one kind of block for another are the result of unusual encounters or stresses, as from radiation. Having once arisen, these new mutant arrangements also become incorporated into the daughter chains that are thereafter reproduced by the original chains and by their progeny-chains. Some, but not most, of these newer-style chains will happen to be more efficient multipliers than before, by virtue of producing helpful by-products, and these will tend to gain the ascendancy in the competition of chains for multiplication and will eventually mutate again in diverse ways. This sequence of mutation and multiplication is the basis for the process of natural selection discovered by Darwin. For those chains naturally survive and multiply even better whose mutations give them more effective means of utilizing surrounding materials and organizing them in the interest of their own multiplication. From that point on, these materials, now deserving of being called living, inevitably develop increasingly complicated structures, in the form of accessory systems called protoplasm, that fit them ever better for waging their continued struggle for existence. This modern view of the origin of life leaves us no place for a spirit that interferes with natural laws, but it brings us down to chemstry.

Turning now from the origin of living things to their further evolution, we note that after the idea that each life-form was created separately had to be abandoned as a result of Darwin's efforts, persistent attempts were made to interpret their evolution from one another as a consequence of willed operations rather than Darwinian natural selection. This implied some kind of spiritual perfecting or adapting principle, either on the part of the living things themselves or else of internal or external powers that guide them. However, the science of genetics, at any rate in countries outside of the Iron Curtain, has become too well based for the interpretation.

For it is found that the mutations in the hereditary material are ultra-minute chemical events that are not pre-adapted to the needs of the organism. Like any changes made blindly in any intricate organization whether it be a watch, a car, a business corporation, or a work of literature, the great majority of these fortuitous alterations results in a less well-organized structure, one that does not do its job as effectively as before. In such cases the descendant organisms carrying the harmful change tend to die out. It is only in very rare cases that a change happens to be of just such a nature as to improve the working of the organism in some way. But these rare changes are just the ones that, after being received by the descendants, tend to bring about their better multiplication. So, step by rare step, throughout the course of many millions of years, a great assemblage of effectively interacting parts is accumulated that has mechanisms for protecting itself and for taking advantage of opportunities of those kinds that have repeatedly arisen in its past history. We then call the organism's reactions "adaptive." Yet they were built up by a process of blind trial and error, in which the occasional beneficial reaction-potentiality tended to be preserved. Thus the present adaptability found in living things by no means results from any foresight or act of will on the part of its hereditary material. It consists of a host of diverse adaptive mechanisms, each of which represents a kind of distillation of a series of accidental successes from out of a great multitude of failures.

Consciousness

The most shining example of an adaptive mechanism is of course to be found in the brain, and more especially in its ability, in higher organisms, to modify its reactions in consequence of its experiences. This we call its capability of learning. As a part of the development of this learning ability, mind, foresight, and will—in a word, consciousness—finally make their debut in evolution. They come as results of evolution, not as its cause. They are aspects of the functioning of a most elaborate organization, that of the central nervous system. Consciousness is affected in the most varied ways, both gross and subtle, when physical or chemical influences are brought to bear upon the brain, whether through surgery, wounds, disease, toxins, drugs, changes in the oxygen supply, hormones or other substances normally in the blood, alterations of blood pressure, temperature or other accessory conditions, direct physical pressure, ultrasonic anesthesia, artificial electric stimulation of the cortex, or stimulation via the normal nerve pathways. To doubt, nowadays, the complete dependence of the mind on brain structure and functioning is to indulge in the most fanciful of wishful thinking. And to postulate, further, that there can be minds apart from or without brains is willful self-deception.

It is obvious that the possession of this marvelous biological organization is of enormous potential advantage in the struggle for existence, and that it was this very advantage that constituted the natural selection that guided the steps in the development of the organization. In consequence, the activities of the brain are normally focused on the promotion of survival on the part of the individual, or at least of his group. This has resulted from the implantation in the brain of inherent inclinations that tend to give it satisfaction in acts conducive to individual or group survival and multiplication.

The inclination to be of service to others of the family or wider group has been an especially important element in the success of man and of some other species and has been developed with much elaboration. Thus, as Darwin pointed out in his *Descent of Man,* human beings have been provided with genetic bases not only for their intellectual capacities but also for their social and moral predispositions. In cultural evolution, later, great superstructures have been reared on these bases.

The general outlines of the evolutionary story have now been so well established that most of the sects that had clung most stubbornly to ancient ways of thought are now attempting reconciliation with these concepts. However, some diehards, though conceding at last that evolution in general has come about by natural selection, still maintain that just at a certain late point in the course of events (when about 99.99 percent of the story up to the present can be reckoned to have elapsed), an immortal soul was suddenly breathed into the Chosen Creatures. Thereat, although there was no discernible break in the process of gradual physical transformation, the ape at once became human in this all-important respect. Moreover, according to this doctrine, there is a similar invisible step in the development of every human embryo.

All such attempts to inject spiritistic intervention into physical and biological events run counter to the findings, previously referred to, that link the operations of the mind to the physicochemical processes in the brain. They seek to perpetuate the ancient animistic superstitions of primitive man. Naively, he thought of spirit and body as two separable entities. He was sure that in his dreams his soul left his body. On this dualistic outlook, the spirit was supposed to be relatively free, but the body was dominated in two ways: first, by its own natural physical laws, and second, by the spirit or spirits that controlled it and that could override these laws.

This dualistic notion is still shockingly prevalent in our own population. For example, among those who regard themselves as especially emancipated from old-fashioned modes of thought are the editors, writers, and readers of so-called "science fiction." Half a century and even a quarter of a century ago this literature was truly in the vanguard of fanciful yet constructive speculation (as in Wells's *The World Set Free, The War in the Air,* and *When the Sleeper Wakes,* or Aldous Huxley's *Brave New World*). However, now that it has been vindicated by atom bombs and space rockets and has thereby become an organ of popular appeal, this literature has for the most part (with notable exceptions as in the works by Arthur C. Clark and Isaac Asimov) lowered its conceptions to those of the mass media. Even in those journals that are best in literary style, the commonest type story is concerned with minds that communicate by means outside the senses or that directly manipulate matter. In other words, old-fashioned magic and fairy tales are here served up in slightly altered terminology. The second and third preferences in these stories are for tales in which man's militarism is extended to other worlds, and those in which he brings to these worlds his present system of private profit. All this seems to be regarded as the epitome of freedom, and as a blazing symbol of what the free world strives for.

Only a small minority of our population ever get into a psychology course in college or even into one in zoology. Yet unless they do, and unless they get into an exceptionally good one, they are unlikely to encounter, to any serious extent, the evidence referred to above, showing that the mind constitutes a great group of functionings of the body, developed through natural selection. They will miss practically all the arguments from physiology, pharmacology, psychiatry, comparative zoology, and genetics, and are likely to remain adherents of the ancient doctrine of dualism.

Indeterminacy

But even those who have been through the mill of scientific evidence are likely to find themselves confronted, at last, with a final contrivance of demonology, reserved for latter-day sophisticates. For in recent decades a new loophole for the insertion of old-fashioned "free will" has appeared from the unexpected quarter of the physicists, in the form of the so-called principle of indeterminancy of Werner Heisenberg. Actually this has nothing to say about consciousness or will, but only concludes that it is not possible, at least by present means, to ascertain both the exact motion and position of an extremely small particle at a given instant, since we interfere with its motion or position in the very act of measuring them. Operationally, then, we are not even justified in concluding that it has a precise motion and position. There is, accordingly, a range of possible issues for any such particles; moreover, a series of them will fulfill these diverse possibilities according to the frequencies that apply to the statistics of randomly distributed events.

Here then is a fine opportunity for the diehard to retreat into the recesses of the infinitesimal, by supposing that a free spirit intervenes to determine just how the tiny particle will move. Moreover, this spirit chooses his interventions so strategically and omnisciently as to pull critical little triggers that set off important events on our own scale in the directions that he chooses. He must also be supposed to intervene so seldom or so astutely as to leave the statistics of the particles still in conformity with the already worked out mathematics of probability.

The gods or demons here at work are such elusive ones that they will probably succeed for a long time in offering these services to the propagandist who is eager to ride on their shoulders. For these riders, the situation is a very convenient one, especially in its relation to the workings of the brain. For now spirits beneath are postulated that act as masters of the brain's physicochemical operations, while they still leave the processes of consciousness of the brain as a whole tied to those operations. Thereby a lower-level dualism results in an upper-level monism that seems to conform with the findings of biology. This is a prime example of the modern apologia.

It seems uncivil to embarrass these apologists with some questions. First, if our minds do depend on the physicochemical workings of our brains, even though the demons can pull wires in these workings, how in turn do the minds of the demons work? Do they have sub-brains manipulated by sub-demons, and so on ad infinitum? In any case, how do the demons in our brain themselves decide that it should take this mental step or that one? Moreover, if these steps do not represent lawful results of previously acting circumstances, how is it that we can often look back on our own decisions and those of others (including animals) and see that they were reached on the basis of already existing inclinations, together with experiences that had indicated to the cognitive faculties how those inclinations might be gratified?

In fact, the more carefully we look into these operations, the more we see these forms of determinism at work. And the more rigorous and comprehensive the reasoning leading up to a decision has been, the more likely is that decision to be in the individual's interests, and to be one that he would insist on as representing his serious choice. Admitting, now, that our seriously considered decisions are based on prior inclinations whose expressions in action are guided by acts of reasoning that have laws of operation, it must next be conceded that both these inclinations and the ability to reason in the given manner are properties of a nervous system supplied through its heredity with certain proclivities and faculties that have been modified by the experiences and learning to which it has previously been subjected. Moreover, regardless of how individual electrons may have behaved, this system as a whole has reacted in a deterministic fashion. And precisely because it has, it is able to give responses that are useful to it.

When we have choices to make that we regard as less important, we are of course more likely to be swayed by minor considerations and inclinations that put themselves forward from among a considerable group of such potential factors and that are correspondingly less predictable. Often in such cases one of our predominant inclinations is that love of variety itself that has been an invaluable inherent asset in primates, by giving them scope for their unusual versatility. Thus we often choose the new or different experience just because it is different. We are then especially likely to attribute our decision to "whim" or "free will." Moreover, we often take self-centered delight in having "our own" way, in making "our own" decisions, and in exercising power. Thus we may in such cases insist that we took a given action "for no reason at all," and may resent the idea that a reason could be found. Yet even here deeper search shows those deterministic factors to have been operating that are rooted in our given combination of biological heritage, cultural milieu, past individual experiences, and immediate situation.

It is by a confusion of ideas, then, that some of us seek to deny causation in the attainment of our decisions. In fact causation of the types mentioned provides us with the only possible kind of freedom: that of making choices in conformity with our natures, choices which, these natures being what they normally are, are likely to bring us further forward in the game of life and to bring us greater fulfillment. If on the contrary we were not to make our decisions in this way, the alternative would be to make them irrelevantly and unpredictably, that is, in *meaningless* ways; why should we desire that sort of chaos? Conversely, the way to increase our only kind of freedom still further is to learn even better the rules of operation: not only the laws of our surrounding physical environment but also those working in our fellow men and within our own inner selves. By doing this we can reach decisions better calculated to attain our aims and can choose aims better in conformity with one another, with the opportunities before us, and with our basic predilections.

Relativity of Freedom

Practically all human beings possess some freedom, in the above sense, to choose our thoughts and our courses of action in our strivings for the greater fulfillment of our natures; nevertheless, people in different circumstances have very different ranges of choice and amounts of opportunity for attaining rewarding experiences. Thus they possess different amounts of freedom. For one thing, they may be hemmed in by very restrictive outer conditions. When, as in the case of primitive peoples, they have never been aware of any other possibilities, they may remain quite content and will consider themselves free. I found the same to be true among many factory workers in Russia who marched together singing a rousing song about how much freer they were than workers anywhere else. This is the pathetic freedom of those born blind. Similarly, men of the future will doubtless pity us for our own present limitations of thought and action.

Not only circumstances external to men but, just as truly, the limitations and disharmonies of their own inner natures set limits to their freedoms. These limits vary enormously according to the richness and the degree of coordination of their natures and according to the backgrounds of knowledge and self-control they have. For the most part, these differences between present-day groups of men are not inborn but result from the type of bringing up, traditions, and behavioral patterns that their families and social groups have supplied them with.

Restrictions on freedom become doubly irksome to a group of people as soon as they realize not only that they may become relatively free of them but also that there are other groups already enjoying this freedom, especially if they regard those groups as blocking their own way to the same freedom. It is in these cases, where resentment can be focused on other persons for supposed deprivation of glimpsed opportunities, that the longing for freedom is likely to be most acute. It is then also most likely to generate social action.

From these considerations, as well as from the Weber-Fechner law of diminishing appreciation with increase in quantity, it follows that when one social class or even one individual attains more freedom at the corresponding expense of others, the total freedom tends to be debased. The maximum freedom, then, does not mean the coralling of as much of the goods and opportunities of life as possible for one's self, regardless of the detriment to others, as seems to be assumed by some devotees of private profit as an absolute good. However, the appropriate organization of society to maximize our freedoms, and our sense and enjoyment of freedom, is a problem that admits no one final solution, as our means of production, transportation, distribution, communication, computation, and education change and improve. This problem is one that we must be forever working on, in as tolerant a spirit as possible, ready to experiment extensively and to hold on to and extend whatever types of organization we find to be good.

Jointly Exercised Freedom

None of these developments would be possible, nor would the culture of humanity have arisen at all, if we had not had the genetic basis for fellow feeling. Men and women will not love one another just because they are exhorted to do so. They could not even be trained to genuine kindliness if they did not, as a result of the type of natural selection that has operated in their past biological evolution, possess the germs, inborn within them, of love and fellow feeling. But since, fortunately, they are made in this way, it is a part of their own fulfillment to exercise these feelings and act upon them. That is, they have the potentialities to pursue these motivations of their own free will, provided that they are given appropriate opportunities. It is a chief function of the family, the school, and the society in which they live to provide such opportunities and thus to encourage the development of socially directed feelings and behavior. These emotions and activities, then, carry for men their own justification and bring their own reward.

In fact, it is in man's nature to experience an especially deep and intense sense of fruition when participating in a great common endeavor, directed toward accomplishments far larger than those that an individual could achieve by his separate efforts. And although the coordination required in such work entails considerable restrictions on his individual courses of action, nevertheless he attains a sense of sharing in the higher freedom of operation and experience on the part of his group that the common endeavor represents. In addition, he finds it rewarding to play his part, no matter how modest, in the judging of the means and goals for the joint project, and in the reaching of decisions concerning them. This is essentially the democratic method.

In addition to constituting its own reward for the participants, such cooperative behavior on their part has far-reaching practical effects that tend to raise them to a higher level of freedom as a result of the material and mental victories that it actually brings. This is, after all, the advantage on the basis of which natural selection has operated in the past in developing the genetic foundation of man's cooperative disposition. The result has been the accumulation of culture which has raised man far above all other ceatures on this earth. Thus, to work together does not, except under the most barbarous circumstances, mean slavery. On the contrary, it means, in its better developed expressions, the highest types of freedom that we can conceive of, freedom of mind, of heart, and of body.

Of the development of our more physical freedoms by these means we have recently had startling illustrations. For it is vividly apparent that our joint endeavors over the ages, capped by modern science and by the unprecedented organized efforts of our own day, are at last delivering to us cosmic forces and can even bring us far beyond our solar system. There, unprecedented opportunities await our urges for new and different experiences and for achievement.

It is true, however, that in the present turbulent transition period, diverse forces of ignorance and fanaticism, not confined to any one group of countries, are threatening to wield men's terrible new powers in the service of their own obsessions, and that in doing so they could wreak unparalleled destruction: of life, of the means of living, of our hard-won cultural attainments, of our values. It behooves us who are aware of these imminent dangers to exert ourselves mightily, by all the means available to us, to prevent the threatened catastrophe. At the same time, we must do so in such a way as to forestall the triumph of those forces, fed on strife, that would curb the most essential of all our individual freedoms—those of our intellectual and moral life.

Nothing is clearer today than that our modern advancement in human control over nature, which is making a better life and more scope for free choice on the part of the average man physically possible, has arisen basically as a consequence of an increasing freedom of men to manipulate materials technically, combined with their increasing freedom to ponder upon and discuss the nature of the world about them and of themselves, theoretically. No sharp distinction is justifiable between these two closely interwoven and interdependent lines of activity, for hand and mind are complementary and are separated only to their detriment. But each major advance made by either involves some break with the old ways, and progress demands that both be as free as possible to explore, test, and build, despite their challenging of interests vested in time-honored usages.

More than that, since men today build on each others' work, modern science, whether physical, biological, or social, demands the utmost freedom and frankness of communication, criticism, and counter-criticism by all engaged in it. The intellect should be subject only to the discipline of thought itself, of intellectual honesty, and—unfortunately but unavoidably—of those self-unrecognized prejudices and ingrained biases to which even the freest minds are liable and which only untrammeled criticism by others can expose. This procedure must be followed in every field, for all are interrelated, and a lagging or retrogression at one point may adversely influence neighboring fields. The issues involved in such progress are not such as should be passed upon by politicians, lawyers, leaders of governments, religions or other ideologies. They can be pursued fruitfully only by those actually engaged in them. Yet in times of stress, like those of today, there is a growing danger of interference with our intellectual freedoms, and attempts increase to dictate our conclusions and direct our researches, on the part of governmental and private organizations, and even the general public.

Notwithstanding today's threats of total dictatorship and total war, scientists must continue working to promote the spirit of intellectual honesty, as well as that of humanity and brotherhood. In the end, though we will all be dead, these spirits will prevail, for their opposites eventually destroy one another. In raising themselves, men will learn that they must conquer not only the atoms and the stars but also the shortcomings within themselves. This work will never be finished, for there is no final "perfection." However, by joining together, men should be able to move on to ever freer positions.

It is to be hoped that we will try in these ways to prove worthy of our descendants. This task involves as one of its major features the effort to produce descendants who are on the whole better, more intelligent, and therefore freer than ourselves. The privilege of striving for this objective constitutes the loftiest of all our freedoms.

The Radiation Danger

"WITH the penetration of science into the world of atomic nuclei, humanity has entered a new epoch." So runs one of the statements drawn up by a group of twenty-four scientists* from ten countries, including three Communist countries, that met in July 1957. They had been called together by Bertrand Russell (whose illness prevented his attendance) at the home of the American industrialist Cyrus Eaton in Pugwash, Nova Scotia. I think that few people would dispute their statement. What is perhaps not so generally realized, however, is the equal or greater importance of another facet of this so-called nuclear age: namely, the penetration of science into the world of cell nuclei, the abode of the chromosomes and genes that constitute our genetic heritage. As Dr. Theodore T. Puck of the University of Colorado Medical Center once remarked to me in private conversation, it is a very fortunate circumstance that the rudiments of men's knowledge about the nuclei of living things, and of the effects of radiation upon them, were gained before they discovered means of unleashing changes in the nuclei of atoms. For men were thereby forewarned of the need for taking the most rigorous precautions, in order to prevent the radiation resulting from the induced changes in atomic nuclei from doing too much damage to their own cell nuclei and thus to the biological basis of future generations.

However, this forewarning has not yet been very effective. For modern knowledge has become so complex and so compartmentalized among specialists of diverse fields that many people who

* Including H. J. Muller.—Ed.

deal with the utilization of atomic nuclear changes and with radiations such as X rays, of essentially the same types as those resulting from atomic nuclear changes, have, until very recently at any rate, failed to take sufficiently into account the already existing knowledge concerning the effects of this radiation on cell nuclei. But beginning in June 1956 a series of pertinent warnings of radiation danger has been issued by groups of scientists called together to consider our state of knowledge on the matter. Included here was one committee meeting under the auspices of our National Academy of Sciences, another under the auspices of the British Medical Research Council, a third convened by the World Health Organization, and a fourth by the UN Scientific Committee on the Effects of Radiation on Man. All of these groups, as well as others representing government-sponsored national and international commissions on radiological protection, proved to be in substantial agreement regarding the danger. As a result of their reports, the importance of taking more precautions than hitherto in operations involving radiation is now at last receiving considerable attention in our governmental and medical circles, and steps are being taken to effect some needed reforms in existing practices. At the same time, large sections of the public also are becoming radiation conscious, although often in a misinformed, disproportionate way, so as to embarrass better informed advocates of radiation protection by their inappropriate distribution of emphasis on different phases of the problem.

Despite the present rise of awareness of this danger, it must be admitted that even today there are persons holding important positions in connection with operations involving radiation who fail to recognize the need for taking the radiation hazard seriously. This situation is illustrated, for example, by a statement made in the *American Weekly* of 2 January 1957, by Dr. Dwight H. Murray, then president of the American Medical Association, reassuring the public that the great majority of medical men are duly cautious in their applications of radiation. Fortunately, there is a growing number of highly qualified radiologists, such as Dr. Raymond R. Lanier of the University of Colorado Medical Center and Dr. A. J. Campbell, Chairman of the Department of Radiology of the Indiana University Medical Center, who take the opposite view.

Another example is provided by an article by John W. Robinson (a medical writer for Science Service!) in the *Rocky Mountain News* of 14 August 1957. This article asserts in its headline that "Fallout Fails to Increase Radiation in Man," but the text of the article reads, instead, that "Fallout . . . has not *dangerously* (my italics) increased radiation in man." The question of course is, what is considered dangerous here? But there is no discussion of this point in the article and no figures are given. It turns out, moreover, that only the radiation in man derived from substances that have entered into his body is taken into consideration, and then only the radiation in substances that later appear in the urine. Anyone knowing anything about the subject knows that such radiation in man is by far the smallest part of that which he receives from fallout.

Still another recent illustration of neglect of the danger is provided by an Associated Press dispatch in *Rocky Mountain News* of 6 August 1957. According to this dispatch, Colonel Barney Oldfield, director of information for the Continental Air Defense Command, headquartered in Colorado Springs, declared that "the man who has taken the greatest atomic radiation jolt and is still alive has fathered two children since the radiation experience and they are healthy, bright youngsters." It is difficult to see what purpose there could have been in the making of this statement unless it was intended to convey the impression that this case provides ground for doubting the conclusion of geneticists that ionizing radiation is damaging to the genetic material. In this case it should also be noted that the disregard of the radiation danger on the part of the authorities concerned apparently went so far that the recipient of the unusually high radiation exposure was not sufficiently warned of the damage that he would be likely to inflict on posterity in case he engaged in reproduction subsequently to the exposure. It is shocking to find such militant ignorance exhibited by persons in whose hands are entrusted decisions that involve our destinies.

Elusiveness of the Damage

As long ago as 1929 I called attention to the fact that one could hardly expect to find evidence of the damage that radiation does to heredity merely by inspecting a number of offspring of individuals that had been heavily irradiated. This had certainly proved to be true in my own experiments with the fruit fly, *Drosophila,* even though in these same experiments it had been possible, by the use of specially designed techniques of breeding, to demonstrate that very significant genetic damage had in fact been inflicted. It was therefore no surprise to geneticists that, later, even the children of the Hiroshima and Nagasaki survivors did not show clear-cut signs of damaged heredity. Yet this lack of definiteness in the Japanese data is, as the investigators of these statistics, J. V. Neel and W. J. Schull, have stated, in no way inconsistent with the existence of considerable hereditary damage in these Japanese populations, damage much greater in amount than would have been expected for *Drosophila* and on a par with that expected for mice. This being the case, the citation of some additional instances of seemingly normal children of irradiated parents in this country can hardly change the judgment on this matter that has been based on more scientific tests.

To date, all types of organisms that could be subjected to rigorous genetic testing, by means of the controlled breeding of both irradiated and nonirradiated individuals derived from relatively pure-bred lines, have provided, when put to such tests, clear evidence of the production, by the radiation, of damaging hereditary changes, that we call mutations. In the course of the past thirty years many types of organisms have been tested in this way. Included in this list are microorganisms of varied kinds, such as bacteria, one-celled animals (paramecia), and diverse types of molds. Included also are lower green plants, such as liverworts, and higher plants, such as barley, maize, the Jimson weed, snapdragons, and cotton. Diverse insects, including not only flies but moths and wasps, and vertebrates, notably mice, have told the same story. There is no reasonable possibility that man is an exception to the rule. Instead it is very probable that, like the mouse, man is unusually susceptible to having his hereditary material damaged by radiation.

There are two chief reasons for the apparent paradox that, on the one hand, the offspring of heavily irradiated parents usually seem to be within the "normal" range of variation of the population, but that, on the other hand, it can be shown by rigorous genetic testing that many of them have incurred hereditary damage. The most direct reason is that the great majority of mutations produce only slight effects on the individual who has inherited the given mutant gene (that is, the changed hereditary element) from only one of his parents, and a corresponding gene of the original, normal type from his other parent. The popular idea that a mutation ordinarily results in a monster or freak is a gross distortion of the facts, although of course there are occasional very rare mutations of that type, both among those produced by radiation and those of natural origin. Moreover, the radiation mutations, as a class, are not more likely than the natural ones to have bizarre or outlandish expressions, contrary to what some writers of sensational stories have depicted.

Now, even though the effect of a mutation is usually slight, it will nearly always be *somewhat* harmful. And since it will tend to be handed down through generation after generation it will finally, in some descendent, happen to find itself in a situation where it acts as the deciding factor in causing that individual's premature death or failure to reproduce. It will thereby bring that line of descent to an end just as surely as if it had exerted a drastically harmful effect on the first individual that inherited it. It is also to be noted that the small mutation, although proving actually disastrous only to the last member of a long line of descent, must in the numerous intermediate generations have slightly hampered, even though usually not recognizably, many of the bearers of it. In these ways a given mutation having only a very slight effect, that is handed down through many generations, tends in the end to do as much *total* harm as a montrous mutation with a drastic effect, such as absence of limbs, that dies out quickly. Hence we are not justified in regarding mutations with small effects as being less important than the more noticeable mutations. As a matter of fact, all of us carry a good many of these small mutations, of natural origin, and we are so appreciably weighed down by the total load of them that it would be reckless to add to this load in our descendants by creating additional mutations.

The second reason why the mutations induced by radiation are seldom to be recognized in the children or in any descendents of irradiated individuals, in populations in which interbreeding of the ordinary hetereogeneous kind is being practiced, is because the interbreeding individuals already differ from one another rather widely in regard to numerous characteristics, as a result of *natural* mutations that had taken place in many previous generations. These mutations of natural origin cause the range of variation that we have come to consider normal among individuals to be so wide as largely to swamp out the effects of any additional changes of small degree that the radiation has produced. In addition, there are usually large variations of a nonhereditary nature, caused by differences in the environments in which the different individuals have developed and are living. Yet all this other variation does not make the superadded damage caused by the radiation-induced mutations less real; it only makes it impossible to recognize them as being products of the radiation.

Although the presence of the mutations produced by radiation is ordinarily obscured by the factors just discussed, nevertheless it is possible to overcome these factors by rigorously controlled breeding experiments and so to bring the mutations to light. It would take far too long to explain here the details of the genetic techniques that must be used. However, it may be stated that the methods are designed to allow a mutant gene to attain heightened expression by being inherited, in the second or third generation after the irradiation, by individuals who do not have at the same time a nonirradiated gene of the original type, derived from the other parent, that would obscure the effect. Secondly, the numerous so-called lethal and sublethal mutations, which, when allowed this heightened expression, would cause the death of all or some of the individuals bearing them before they could be found by the observer, are made detectable through the change in ratio of certain visible characters among the offspring of an affected family. For this purpose certain predetermined crosses are necessary. Thirdly, the potentialities of the stock used in these crosses must be thoroughly known, and they must have been bred previously in such a way as to reduce to a minimum the number of mutant genes preexisting in them that could give rise to effects confusable with those of the mutations produced by the radiation. Finally, the environmental conditions must be well enough controlled to keep at a relatively low level the occurrence of noninheritable variations, that is, acquired characters, of environmental origin. None of these rigorous requirements can be met in any observations on irradiated human populations.

The Core of Our Being

It may be well at this point to go back and supply background information, most of it long known, concerning this hereditary material that we have been talking about. It should first be recalled that the hereditary endowment of a person, or animal, or plant, is in the form of minute threadlike bodies, the chromosomes, each of which, like a necklace, is composed of numerous smaller parts, the hereditary particles or genes, that are of diverse types and are strung together in a long thin line. Each cell of your body contains your full equipment of chromosomes and genes, in fact, two sets, one set derived from your mother and the other from your father. Before a cell undergoes another division into two, each gene in it gathers appropriate material from the intracellular fluid about it, so as to piece together, next to itself, a new gene with a pattern like its own; that is, it reproduces itself. New light has recently been thrown, by J. D. Watson and F. H. C. Crick, on the way in which this remarkable process of gene reproduction occurs. In consequence of the simultaneous reproduction of each gene in each chromosome thread, there are now two lines of genes, that is, two chromosome threads, lying parallel to each other, in the place of each original chromosome thread. The threads then coil into tight spirals, thereby forming the familiar sausage-shaped bodies (often visibly double) that the word chromosome is ordinarily applied to, and that are to be seen under the microscope at the time the cell divides. Then, during cell division, one member of each chromosome doublet is pulled to one end of the dividing cell while the other member is pulled to the other end. Thus the group of chromosomes gathered at each end, representing the nucleus of each daughter cell, comes to have an outfit of chromosomes and genes just like that which was in the original mother cell.

A human being's hereditary endowment consists, according to

some recent work by J. H. Tjio and A. Levan and by C. E. Ford and J. L. Hamerton, of forty-six chromosomes, one set of twenty-three derived from his mother and a corresponding set of twenty-three from his father. Each chromosome contains hundreds or thousands of genes, all chemically different from one another, so that a whole set of twenty-three chromosomes probably contains at least ten thousand different genes or, according to some estimates, as many as fifty thousand. Each gene has its own distinctive, precisely specialized, role to play in the complex web of biochemical reactions that so amazingly cause the egg to develop into the mature individual, with its multitude of nicely interacting parts, and in the biochemical reactions that operate, further, in the maintenance of that individual, and at last in his senescence.

It should also be realized that any one of these genes is itself a highly complex body, containing a string of tens of thousands of more elementary parts, called nucleotides, that are of four types. We believe that it is the exact sequence of these four types of nucleotides in line within the gene that determines just what capabilities of reacting that gene has, even as the arrangement of letters in line in a word determines the significance of that word. A mutation, judging by present evidence, consists in a sudden, permanent alteration in arrangement of one or more nucleotides within a gene either by substitution, addition, or loss, and this alteration usually results in some change, often slight, in the reaction potentialities of that gene.

The exact arrangement of nucleotides in any given gene has been arrived at through billions of years of trial of one accidentally occurring mutation after another. The great majority of these naturally occurring mutations did not result from radiation but from the accidental encounter of the gene with a chemical situation, such as the strategically placed impact of an oxygen atom, that broke one or more of its distinctive chemical bondings and thereby allowed other bondings to be made. The mutant gene resulting from such an accident is in most cases less well fitted than previously for carrying on the biochemical work required of it, and in these cases the individuals inheriting the mutant gene tend to die out. Everyone knows that a precise machine is more easily injured in its workings than improved; that is, most changes made at random are harmful to it. But in the very rare event that the alteration happens to result in a mutant gene that does its job better, or that does a somewhat different but even more useful job, the individuals inheriting that gene will tend to survive and multiply in the struggle for existence. The altered gene in the descendants will again change in different ways and the few individuals having a second advantageous alteration, in addition to the first, will again survive preferentially. Thus in the course of billions of years a whole succession of advantageous mutations will gradually become accumulated in any given gene, making its present structure one that is most precisely and complexly adapted for performing its specialized functions.

We are thus confronted with the seeming contradiction that mutations have furnished the building blocks, so to speak, of evolution, but that nevertheless more than 99 percent of mutations are harmful, tending to cause some kind of impairment of function. Mutations produced by radiation are not, as a class, different from naturally occurring mutations in this or other respects. But it remains true that if we expose a group of animals or plants to radiation and thus increase the total number of mutations that occur in them, we can, by applying specialized means of searching, find more mutations of advantageous kinds than we do in untreated populations. By then breeding selectively from the individuals having these fortunate changes and crossing them together, we may actually improve the breed in regard to one or more characteristics that we consider desirable. So for example, A. Hollaender, M. Demerec, and their coworkers, by repeatedly irradiating the mold *Penicillium* and each time practicing drastic selection, have obtained lines giving a far higher yield of the valuable antibiotic penicillin. But it must be emphasized that this result has been achieved only by the sacrificing of the hundreds of defective mutants, not to speak of all the nonmutated individuals that were discarded, for every case of one with increased yield that was bred further. If we consider other species, we see that the more expensive the individual specimens of a given species are to raise, and the more important they are in themselves, the less practicable becomes this wasteful method of hastening evolution in that species. For man, it would be quite out of the question.

Our Burden of Mutations

To understand this point more clearly, it should be remembered that the mutations of natural chemical origin—the so-called spontaneous mutations—are occurring in every generation with a fairly high frequency, and that the vast majority of these are harmful. It can be estimated that at least one person in five contains a new spontaneous mutation, one that arose in one of his parents. In addition, every one contains dozens of these mutations that arose in still earlier generations and that have not yet met genetic extinction by causing the death or failure to reproduce of a person containing them. These genetic extinctions must, on the average, be about as numerous, per generation, as the cases of new mutations that are arising. For if they were less numerous the number of mutant genes present would rise in each generation until the population became heavily enough loaded with defects so that extinction *would* occur at a rate as high as that at which the new mutations were occurring. This rate according to present conservative estimates is about one in five, that is, about one person in five must meet genetic extinction. These extinctions, however, are only one expression of the load of mutations. They represent the least fortunate cases. The general population has accumulated from scores of past generations so many mutations, each one of them usually having only a slight effect, that the total load of them is distributed fairly evenly among different people. Thus, with one-fifth of the individuals loaded down enough to meet extinction, the average individual may be said to have a total load great enough to give him a chance of extinction of about 20 percent, and the great majority of people would carry loads fairly close to this average.

In consequence of this, each one of us carries his own private burden of slight disabilities, such as a tendency to allergies, or colds, or pains in the joints, or astigmatism, or difficulty with mathematics, or nervousness, or high or low blood pressure, or need for extra amounts of some vitamin, and sometimes also one or more somewhat greater disabilities, such as an unusually high tendency to cancer, or to developing a psychosis.

The successes of modern medicine make these weaknesses much less evident and burdensome nowadays. However, we must remember in this connection the law of mutational equilibrium previously mentioned: that is, that mutant genes will tend to accumulate generation after generation until about as many genetic extinctions are occurring per generation as the number of new mutations that are arising. Thus, the amelioration of our present biological burdens by modern medicine, sanitation, and all the other artificial aids to living is only temporary; since these devices save for reproduction many of the more defective individuals who would otherwise have died off, and they will continue to operate in this manner until, many generations later, the average individual, even with the best medical care and man-made supports, will have accumulated so heavy a load that he is just as likely to meet genetic extinction as was the man of the stone age. Moreover, he will be as much dragged down by his natural disabilities as that man was. The only way to avoid this anticlimax would be for the people bearing heavier genetic loads to refrain from reproducing, even though medical and other aids had made them able to live and reproduce. We are at present a long way off from such social motivation in reproduction.

Now it is evident from all this that if the frequency of mutations is increased, as by exposure to radiation, there will be a resultant increase in the mutational load. Each added mutation will mean, on the average, an additional case of genetic extinction, usually occurring far in the future, together with slight added handicaps for the line of individuals living in the generations prior to extinction. If, for example, enough radiation is received by a population to produce as many mutations as occur naturally per generation, and if this radiation continues to be received generation after generation indefinitely, the mutational load of that population will eventually, after one to a few thousand years, rise to twice its natural load, and so will the frequency of genetic extinction. If however this much radiation is received by only one generation, say the present one, the load will be raised by only one to a few percent, since the mutations of any one generation comprise only one to a few percent of the total load that is carried. But, as if to make up for the fact that the addition to the load of the next generation would be relatively small in this case, this extra load would tend to be handed down for scores of generations, incommoding each one in turn, until at last it had subsided by reason of the extinctions it had caused.

It is beside the point to argue that, among hundreds of slight afflictions, there would probably be an occasional case of a helpful mutation that raised the possessor's ability in some respect, for the extra burden would far outweigh the rare benefit. Why should we pay this exorbitant price for the supposed benefit, when beneficial mutations of comparable types are arising naturally in the population anyway and could be found in sufficient abundance if we looked for them? Would it not be far saner if these more fortunate individuals, these natural beneficiaries, were enabled, by favorable social arrangements, to have more than the average number of descendents, rather than that hundreds should be weakened by radiation in order that one person might by chance be benefitted.

And if this more fortunate mutant were produced by radiation, what guarantee would we have that with our present ways of living this one person would really succeed in leaving a larger number of children? It is in fact contended by some students of the subject that modern economic and social conditions lead rather to a faster multiplication on the part of those with lower abilities. In such a situation any increase in the frequency of mutations could not hasten evolution but would only accelerate the decline. If, however, we were able to reform our social arrangements or our mores in such wise as to bring about the opposite trend, one that led to a higher rate of reproduction on the part of those who were naturally better equipped with those capabilities that are universally acknowledged to be desirable in man, then that would be a way of raising the genetic level that did not require the simultaneous increase of all our biological burdens that a rise in mutation rate would entail.

Magnitude of the Radiation Damage to Posterity

Let us now try to form an idea of the approximate amount of hereditary damage that a given amount of radiation will do. It is necessary first to mention that the amount, or dose, of radiation received from radioactive substances or from X-ray machines is commonly expressed in terms of roentgen units, denoted with the letter r. It may not mean much to some readers to be told that lr produces about two billion pairs of ions per cc. of air and some seven hundred times as many in water or tissue. Perhaps it would mean more to say that about 500r delivered to the whole body in a short time, say half an hour, gives rise to such intense radiation sickness as to kill half the people exposed to it within a few weeks, and that it is therefore called the semilethal dose. It is also likely to cause, after an interval of some weeks, prolonged sterility in the survivors, even to persons who have had only their reproductive organs exposed to this dose. One fairly prolonged fluoroscopic X-ray examination of the lower portion of the trunk may deliver several r to the reproductive cells, unless they have been specially protected, as by a lead screen or by shuttering down the beam. X-ray photography usually gives a much lower dose to the reproductive organs, but pelvimetry, for example, generally gives 4–5r to the reproductive organs of *both* mother and unborn child.

It is reckoned that in the United States a person on the average receives nearly 5r or possibly more in his reproductive cells from medical and dental diagnoses and treatments before the time at which he produces a child. If pelvimetry by X rays as now done becomes routine, the average amount would be two to four times this. The amount from natural sources of radiation also averages nearly 4r for a thirty-year period. That derived and to be derived from fallout from the nuclear tests held to date may add up to perhaps a twentieth of an r.

How many mutations will such doses produce? As to this, experiments with diverse species are in satisfactory agreement in showing that in ordinary cases the number of mutations produced in the genes of the reproductive cells is proportional to the total amount of radiation received by them, no matter whether the delivery of that radiation has been concentrated into a second or less or spread out evenly in dilute form over many years, or given in the form of many tiny exposures at widely separated intervals. For instance, twice a given dose results in twice as many mutations, while one hundredth the dose results in a hundredth as many mutations as the given dose itself produces. Thus there is no threshold dose, *no dose so small that it fails to produce some genetic effect, or some risk of effect,* proportional to its size, and all doses received at all times prior to reproduction sum up accurately in deciding the total number of mutations produced in the genes of the reproductive cells.

There are several methods, all involving a considerable range of error and the carrying over of results from other animals to man, for estimating the number of mutations produced in man by a given amount of radiation. William Russell's meticulous experiments at Oak Ridge with individual genes of the mouse have shown that it takes about 35r to produce as many mutations as arise naturally in one generation in that animal, that is, 35r may be called the dose that doubles the mutation frequency. Russell finds that if we concentrate attention on one given gene, we find that in one reproductive cell in one hundred thousand that gene has had a natural mutation in it that was not present in the reproductive cells that produced the previous generation. To use this figure to find how many mutations are arising in the whole set of genes, we should have to know how many genes there are in the set. As to that, the number ten thousand is a very conservative estimate, based on the fact that there are found to be some ten thousand genes in the whole set of the fruit fly and that there are several reasons for inferring the mouse to have even more genes than the fly. It can therefore be reckoned that the chance of a natural mutation arising in some gene or other of a reproductive cell of a mouse is at least one mutation in every ten reproductive cells. Since each offspring is derived by the union of two reproductive cells, egg and sperm, the chance of the offspring receiving a natural mutation becomes at least one in five.

Now since 35r would produce as many mutations as arise naturally in one generation, the offspring from parents both of whom had received 35r would on the average contain one newly arisen natural mutation and in addition one artificially induced mutation among every five individuals. On the basis of this result we see that in order to have, on the average, one *radiation-induced* mutation in *each* offspring, both of the parents would have to have received 5 × 35r, or 175r. The same result would be produced if just one of them, say the father, had received 350r and the other none at all.

In man, there is some evidence, from more than one direction, indicating that the natural mutation frequency is similar to that in the mouse but probably somewhat higher, perhaps twice as high. As for the frequency with which mutations are produced by radiation in man, the little evidence available admits a wide range of values, well within which lies the frequency found for the mouse. Since results for other mammals have not yet been obtained, the mouse frequency must in the meantime be used as our safest guide. Our provisional estimate, then, is that it takes some 350r, or less, given to one parent, or 175r to both, to produce an average of one mutation per offspring in addition to those already present. Thus, in a population containing one hundred million people who had not yet passed their period of reproduction, and who would on the average still reproduce a number equal to their own, the application of 175r (a dose probably somewhat greater than the average received by the Hiroshima survivors) would cause the next generation of one hundred million to contain one hundred million newly induced mutations (or one per person) in addition to all their natural ones.

Although this additional load would usually be small in relation to the natural one, and not easily demonstrable, nevertheless it would tend to be transmitted until it had caused nearly this same number, namely, one hundred million cases of premature death or failure to reproduce. However, these one hundred million casualties would of course be spread out over hundreds and even thousands of years, at the rate, at first, of some two million per generation, until at last these induced mutations had thereby been weeded out. This reckoning assumes that the population remained stable in numbers; if it had doubled, the number of casualties would of course have doubled along with it. Even more genetic damage than this might be produced in the case of a large-scale nuclear war, since several hundred millions would be likely to be heavily irradiated. However, since people who had incurred much more radiation than here assumed would be sterilized by it, it is unlikely that the genetic damage to the survivors' descendents would be more than two or three times as intense as this.

By using the rule of the proportionality of the effect to dose, we can similarly reckon the future casualties ensuing from any other dose on the assumption that 175r per person results in one casualty among the descendents for each present parent. So, for example, the estimated 5r received per person before reproduction from medical X rays would in our reproductive population of nearly one hundred million give rise to nearly three million, inherited by the next generation, and therefore, about this many eventual casualties, when all future generations are considered. And if not one but, say, ten generations in succession were all subjected to this much medical exposure, the total effect would of course be ten times as great, or thirty million, and the effect in *each* generation, after the tenth, would be nearly ten times as great as if just one generation had been irradiated.

Turning now to the fallout from the nuclear test explosions already held, and using the previously mentioned estimate of 1/20r per person, which is about one hundredth as much as that received in the United States from medical X rays, we can reckon that the American reproducing population of nearly one hundred million would in the next generation inherit some thirty thousand mutations that had been induced by the fallout, and approximately this many future casualties from that cause, distributed at the rate of some six hundred per generation in the earlier generations, with the number gradually decreasing to nearly zero in about three thousand years. Again, we must note that with every increase in population in the future, all these numbers of casualties would undergo a proportionate increase. It is also to be remembered that the fallout damage is not restricted to the United States but is global, and that the total number of genetic casualties produced by it in the future would therefore be some seventeen times that for the United States alone, that is, some five hundred thousand, without taking into account future increases in population. Of course, if an equal amount of fallout were to be produced by test explosions yet to be set off, all these estimates of the effect would again be doubled.

Damage to the Exposed Individual

Besides producing these effects on future generations by causing mutations in the genes of the reproductive cells, radiation also damages the individuals who are themselves exposed to it, that is, it damages the somatic cells. Not only do massive concentrated exposures, of 200r or more, produce the now widely known symptoms of radiation sickness, which may prove fatal, but there are much longer-delayed symptoms, just as objectionable in their own way, that may result not only from large doses but sometimes also from small ones. There is increasing evidence for the inference, though it is not yet rigorously proved, that most if not all of the effects on the exposed individual himself, whether appearing hours, days, weeks, years, or decades after the irradiation, are fundamentally similar to the effects on the later generations, in that they result from damage done to the genetic material, that is, the genes and chromosomes, but in this case those of the body's own cells, the somatic cells. Time does not permit the citation here of the diverse facts that point toward this conclusion. The damage appears to be of two main kinds: first, that arising from the production of mutations in genes, which may result in diverse kinds of leukemia and cancer, and second, that arising from the production of breaks in chromosomes, that may result in the symptoms of radiation sickness, as well as in long-delayed effects such as cataracts, chronic anemias, and, most important of all, a lessened resistance to almost all kinds of diseases and afflictions, similar to what is found with advancing age.

If these injuries to the genetic material of somatic cells arise at random, point-wise, now in this cell and now in that in the manner in which they are known to occur in germ cells, and thereby form the basis of the symptoms found in the exposed individual, then it would not be surprising if the long-delayed effects, like the germinal mutations, appeared with a frequency proportional to the total amount of radiation that had been delivered in small amounts over a long time. Evidence indicating that this is probably true of the induced leukemias has recently been brought together by E. B. Lewis at the California Institute of Technology and by W. M. Court Brown in England. Evidence that it is true of the reduction of the life span had been obtained over a decade before that by R. D. Boche and others in studies on diverse kinds of mammals and has more recently been amplified by data on man, notably by Harden Jones of the University of California. If we accept their calculations, we are enabled to get rough estimates of the reduction in average life span and also of the frequency of induced leukemia in a group of people with any given exposure.

Except in the special case of fallout, to be considered later, the main radiation damage to the exposed individual himself consists in the shortening of his life span, which is really an expression of a rise of his mortality rate at all ages from miscellaneous causes, not distinguishable from those operating ordinarily. There is approximately .02 percent shortening of life span for every roentgen unit received by the whole body. This in human beings amounts to about five days. However the amount varies with the age at which the irradiation occurs. We see then that if the 5r of medical radiation received on the average by the reproductive organs of Americans in thirty years could be taken as applying to the whole body, which would thereby receive about 11r in a life of seventy years, the average shortening of life from this cause would turn out to be about two months. Actually, however, most of the body would receive more radiation than the reproductive organs do, since these organs are usually further from the direct line of the X-ray beam, and therefore the actual shortening of life would be correspondingly greater. It must of course be remembered that this effect is not an even one that curtails most people's lives by approximately the average amount of two months. Since it consists really in a higher mortality at all ages after irradiation, its incidence is erratic. Therefore, while many people will in such a case not have their life span reduced at all, others will pay the balance due by losing more years of their lives.

In comparison with the shortening of life of the general public calculated to be caused by medical X rays, that from global fallout from tests already held, being only about a hundredth as great as this, is practically negligible. However, it is far from negligible for people who have received the much higher doses given by local fallout, as did some of the Marshall Islanders and Japanese fishermen. Nuclear war itself would of course result in even greater reduction of the lives of those who survived the more immediate effects. This has been strikingly illustrated at Hiroshima, where the mortality among the so-called survivors is perhaps double that expected for any given age group of unexposed persons. This represents an average shortening of life of nearly a decade. Here the effect was intensified by reason of the radiation having been received in one concentrated, intense dose, and also by its sometimes having included neutrons. The intensification of effect here receives a probable interpretation in the fact that under these circumstances there is relatively more chromosome damage than there would be otherwise. This is caused by entanglements of chromosomes, resulting from two or more of them having been broken at almost the same time, within reach of one another.

We may turn next to effects on persons occupationally exposed to radiation, such as medical and dental workers, and the rapidly growing body of persons, said to number some eight hundred thousand already in this country, who are working with industrial, military, and scientific uses of nuclear energy. Any such person who worked for forty years, say from the ages of twenty-five to sixty-five, under conditions that would expose him to the recently promulgated "maximum permissible dose" for such workers, namely, to 50r every ten years, would receive a total of 200r and would thereby suffer an estimated curtailment of life of, on the average, between two and three years. Until a few months ago, the "maximum permissible dose" had been set at three times as high as this presently accepted value.

It should be recognized, however, that in our governmental atomic installations the dose actually received by the workers is in the great majority of cases far below the limit of 50r per decade now set for occupationally exposed workers. But it can hardly be expected that private atomic plants, or even governmental ones in underdeveloped countries where every penny counts, will go to the greatly added expense of providing a great deal more radiation protection than has been officially agreed upon as necessary. In view of this situation, an even further reduction in the so-called "maximum permissible dose" recommended by the committee dealing with this matter is to be hoped for during these formative years of the development of nuclear energy for peacetime uses.

It is also to be hoped that the medical and dental professions and the numerous others, such as chiropractors and shoe salesmen, who deal with X rays, will engage ever more earnestly in the provision of better protection not only for their patients or customers, but also for themselves and their technicians, including, for example, dental hygienists and even receptionists. Measurements have shown that many of these workers have been receiving a far higher dose of radiation than that presently accepted as the maximum permissible one for occupationally exposed personnel.

In passing, we may note that the recommended maximums for occupationally exposed workers should not be confused with the recommendation of our National Academy of Sciences committees and of the Radiation Protection committees that the dose of man-made radiation received by the population as a whole should be kept down to an *average* of 10r in thirty years. For this average would still allow workers who are occupationally subject to radiation to receive the fifteen times higher dose of 50r every ten years, provided that their additional exposures were not sufficient to raise the average for the entire population above the limit of 10r in thirty years. At the same time, a considerable proportion of students of the subject, including members of the committees mentioned, feel that even the present recommendation of a population average of not more than 10r in thirty years is very much on the high side, not so much because of the possible three-months' average reduction of life span it induces but because of the far greater damage it allows to be inflicted on posterity. The figure 10r was adopted as more or less of a compromise with the trend of existing practices. But it was accompanied by another recommendation, that the situation be periodically revalued. This was largely with the aim in view of attaining progressively lower limits whenever they became practicable.

The Action of Strontium 90

In dealing with the effects on the exposed individual himself, we reached the conclusion that the general rise in death rate from all causes, expressed as reduction of life span, represented by far the major portion of the somatic damage. Its total amount, moreover, is much greater than that of any subdivision of it, consisting in the deaths of some particular type. We must however make an exception, in this regard, of the somatic effects produced by fallout. *The generalized* effect of fallout on length of life is due mainly to gamma radiation that penetrates the body from outside, derived from radioactive materials in the environment. This is the only fallout radiation that we have taken into account in our discussions so far and, as we have seen, it amounts in the United States to only about a hundredth as much as that from medical sources. But in addition to this, there is the radiation that is derived from fallout material that people take into their bodies with their food, water, and air. Although it has been estimated that this radiation of internal origin is for the most part even smaller in amount than that penetrating the body from outside, nevertheless there is an element, Strontium 90, that becomes selectively deposited in the bones, within which, and in the immediate neighborhood of which, its short beta rays remain concentrated. The strontium is too far from the reproductive organs for these rays to reach them and damage posterity to a noteworthy degree, and these rays are also insufficient in total amount to give much generalized shortening of life. However, being concentrated in the bones and therefore also in the immediate vicinity of the bone marrow where the blood-forming cells are located, it is to be expected that the beta rays from strontium would give rise to bone cancers and more especially to cancers of the blood cells, notably leukemias. Often these malignancies do not arise until many years after the strontium has penetrated the system.

If, now, we follow Lewis and Brown in accepting the rule that the frequency of the induced leukemias will be proportional to the dose of radiation received and use their figures on the frequency of such cases induced by a given dose, the considerations of the amount of Strontium 90 to be deposited in bone as a result of the nuclear tests already held lead to the conclusion that there would probably be several hundred thousand cases of leukemia induced by this agent throughout the world within the life span of persons now infants. This is a number comparable with the number of casualties that we previously estimated would be produced among all future generations by the radiation, derived from the same tests, that penetrates to the reproductive cells from fall-out materials located outside the body.

Of course it is not intended that any of these figures be taken literally, for they may easily be as much as three times too large or too small. However, they should indicate the order of magnitude of these quantities and also give an idea of their sizes relative to one another, in cases where these sizes differ widely. Thus it is possible to decide that the radiation from fallout, unlike that from most other sources, probably does about as much damage to the directly exposed generation as to posterity. Even at that, however, it must be doing far less damage to this generation than medical radiation as presently applied does.

The Reform of Radiation Practices

All this discussion so far has been largely confined to consideration of the biological effects of present-day practices involving radiation. However, the decision of what to do about these practices involves us in a much wider range of problems. Misleading conclusions might be drawn if the discussion were closed now, without very brief reference to these other matters.

First, as regards that artificial source of radiation from which by far the greatest portion of present-day human exposure is derived, namely the medical X-ray machine, the other problems involved in decisions regarding these practices are of course the medical ones of securing effective diagnoses and treatments. Many medical and dental practitioners declare firmly that the good done by their utilization of radiation far outweighs any damage. This is undoubtedly true in very many cases, while in some it is debatable and in others, such as the induction of ovulation by doses of some

250r applied to the ovary, it involves a risk that is demonstrably greater than the gain.

But our case does not rest on this basis, for we may freely admit that, on the whole, the medical and dental benefits of X rays are enormous. The point is that, in addition to the benefits, damage to our own and future generations is also being incurred, and that by means of suitable precautions, including accurate control of the voltage and total dose, shuttering down of the beam, use of more sensitive films, protection of the reproductive organs by a lead shield or apron, maintenance of a record of the exposures of each individual, omission of unnecessary irradiations, and utilization of other methods of diagnosis or treatment instead of irradiation where feasible, not to speak of the coming technique of photoelectric image amplification, it is possible to achieve just as effective results while delivering to the patient's body, and more especially to his reproductive organs, only a small fraction of the radiation now received. The same holds true, moreover, for the operators themselves. Thus the way is clear for this reform, and it can come as quickly as the natural human obstinacy that has blocked its adoption for the past thirty years will allow. Pressure from the public can hasten this process of reform (even though this pressure is resented), just as it has been instrumental in initiating it.

As to the damage done by fallout, even though it is *relatively* small in comparison with that from medical X rays and from other controllable causes, nevertheless it is, as we have seen, very large when viewed in terms of absolute numbers of casualties, and it is therefore quite legitimately a matter of considerable concern. However, we may well ask, why is the apprehension which it arouses so out of proportion to that aroused by the far greater damage from medical radiation? The answer of course is in the connection of the nuclear tests with the possibility of nuclear war. Here, however, we again step on disputatious ground. Just as medical men have argued that the radiation damage done by them has been justified by the greater benefits received, so many persons concerned with international affairs maintain that the tests have helped in the avoidance rather than in the provocation of global war. If this were true, then the damage done by the tests would be fully justified, inasmuch as global war in our times would be likely to result in thousands of times as much damage as that done by the fallout from the tests.

The Avoidance of Nuclear War

This is not the occasion to discuss the details of these matters, but we cannot ignore them if we would keep our problem of fallout from tests in due perspective. Personally I am convinced that Leo Szilard and Albert Einstein were supremely right in persuading President Franklin D. Roosevelt to support, at enormous cost, the development of the atomic pile and bomb, in view of the likelihood that Hitler might otherwise have had it before we did. Similarly, when plausible reports reached this country in 1949 that the USSR was already developing the hydrogen bomb, it was, so I believe, obligatory on us to get to work on that also. Moreover, tests were an integral part of both these developments.

Today, however, since both sides have these weapons, no nation can afford to initiate a global war. For that would mean, for *both* sides, ruin on a far vaster scale than any in previous history. There is evidence that the leaders on both sides now realize this. It is also true that any major war today would as it progressed be very likely to become a global thermonuclear war, no matter how vehemently nuclear weapons had been banned. For the peacetime reserves of nuclear material are readily converted into warheads, and a nation facing the defeat of all it stands for will use all means at its disposal. There is, however, more danger that a minor war, leading to a major war, will start if nuclear weapons have been officially renounced, and if they therefore present a less evident threat. Once started, the little fuse that had been lighted would lead in the direction of the H bomb. This means that there is no use in pretending that nuclear weapons of whatever magnitude can be put under lock and key by any fiat. Instead, the world will have to adjust itself to them realistically by its own reformation.

Just as gunpowder came to stay and helped to end the feudal system, so nuclear weapons—in this case, however, not by their actual use but by their potential destructiveness—must force this shrinking world gradually to abandon the nationalist system, and ultimately too to discard those fanatical, dogmatic ideologies which divide men into warring political, economic, racial, and religious sects and thereby foster irrational animosities. However, slavery can never become the basis for a secure and desirable peace. Even a return to savagery is hardly a more repulsive prospect than a technically advanced slave state embracing the whole world. No, the only sound basis for world reconciliation is that of intellectual freedom, the scientific world-view, recognition of the brotherhood of all men, and of the right of the common man everywhere to a dignified life and a significant voice in the determination of his own destiny. If we stand firm for these principles, and *live* them in our relations with each other and with other peoples, while retaining our sheathed nuclear weapons as a deterrent to aggression, there are clear signs that other peoples too will gradually accept this viewpoint and that the nuclear weapons will thereby finally become obsolete.

Yet this does not mean that we need, any longer, to sharpen our nuclear weapons further or to pile up more stores of them. A greater offensive power on one side than on the other is meaningless so long as both sides already have the means of totally destroying each other anyway. Further nuclear tests under such circumstances are pointless, and only tend to aggravate animosities, while at the same time wreaking more of their insidious human damage, mostly on peoples who have had no voice in the matter. At this stage, therefore, the abandonment of the tests by both sides is clearly a procedure that is in line with the more general policy of reaching as many compromise agreements as possible, of disarming more and more, on both sides, in regard to conventional weapons and armies, and of furthering all positive measures of economic, political, and cultural cooperation that will tend to hasten the coming of genuine international amity and, finally, world federation.

Nurturing Our Genetic Heritage

Among these positive measures, of course, is the intensive development of the peacetime applications of nuclear energy, including the development of nuclear power for the bringing of a high standard of living to all peoples and the consequent reduction of their mutual jealousies. This, however, brings us up against the final radiation danger, that of the peacetime uses. Unless they are rigorously controlled, both nationally and internationally, these growing industrial atomic uses can readily lead to a situation in which there is an even higher per capita exposure to radiation from this source, throughout the world, than there is today in this country from medical sources. For instance, the problem of the disposal of atomic wastes has not yet been satisfactorily met, in terms that would be suitable when the already projected peacetime nuclear developments have come to fruition. We must insist that a sufficient portion of the energy supplied by the atom nucleus be set aside to enable it to dig an adequate grave for its own wornout slugs and remnants and also that enough of its energy be used to provide sufficient protection to man from the operations of its working parts. Then only can we be assured that we are the masters, not the victims, of the atom nucleus.

All this means that we have to become increasingly conscious of our own cell nuclei, of that genetic heritage which is our most precious trust for future generations. Perhaps the greatest gift to man that nuclear energy can bring is the heightening of this realization, brought about in the first place by fear. With the growth of this awareness, however, we will not be satisfied with just protecting our genetic heritage from the radiation danger. We will also, on looking about us, discern that it is subject to other dangers, that may be just as grave. We may for instance find that some of the chemicals to which we are subjected in our modern life cause as many or more mutations than medical radiation does today. And we may find that some of our social arrangements lead to such a high rate of perpetuation of harmful genes, or to such a strong tendency for genetic extinction on the part of especially valuable ones, that even more deterioration is caused in this way than by improperly controlled radiation. In other words, we, the public, must become genetics conscious in all matters concerned with the shaping of our society, our education, and our personal motivations.

Finally, with this more enlightened outlook, it should become evident to us that the chief springs of action of our present times, based as they are on insecurity and "quiet desperation," and, in matters of genetics, on the fear of biological deterioration, can, in a world that has virtually unlimited sources of power and a realization of the inspiring potentialities of man, turn from fear to hope, and from activities chiefly directed toward avoiding destruction to those which instead bring advancement and improvement. In matters of genetics, this means that we shall not be content with merely protecting our biological heritage from damage but will find our greatest fulfillment in working to improve its quality and to bring into being a humankind formed of an ever nobler fiber. Perchance some day, in working toward this end, radiation may at last prove to serve a useful genetic purpose for mankind. If so, this might be in connection with ultrafine operations, involving the genetic material. But it would certainly not be used in a broadcast, random manner, whereby, for every instance in which it conferred a benefit, it left behind the wreckage of a thousand wasted lives.

In Search of Peace

TODAY, for the first time in history, the offensive power in war has so outrun the means of defense that either one of the two great military coalitions that are now facing each other could practically wipe out the other one. Neither could effectively defend itself nor destroy the other's offensive. Both sides would therefore lose completely. Even neutrals and generations far in the future would suffer significant damage. No foreseeable innovations in means of offense or defense are likely to alter this situation.

The collapse of civilization which modern war would bring cannot be avoided by restricting nuclear bombs to so-called "small" ones, used only on military targets. For as both sides played at this game, the range of destruction sought would tend to grow indefinitely. And even if all nuclear weapons had been renounced to begin with by both sides, the progress of a major war, with its relaxation of international commitments, would see the peacetime nuclear-reactor industries rapidly converting their potentials into nuclear arms. Moreover, modern biological and chemical means of mass destruction, although seldom publicized, should never be forgotten. Easily concealed and transported, these subtlest of weapons might prove as deadly in the end as nuclear bombs.

Today, even a minor war may serve as the fuse that leads to the loosing of these diverse weapons of mass destruction. Thus the whole world is endangered whenever one country embarks on a course of violence, no matter how just its claims might be. Therefore *all* war must hereafter be avoided. Until a more brotherly world is established, negotiation is imperative, if only for the sake of self-interest. At the same time, the arms race must be acknowledged to be not merely futile, in that it fails to lead to a greater measure of security for either side. But it must also be recognized that it intensifies the danger of the conflagration that it seeks to avoid, and that it squanders the efforts and resources so badly needed for construction. If both sides come to realize this, they will see that they must substitute for the arms race its opposite, that is, progressive disarmament. However, disarmament can be agreed to only if it involves a balanced, controlled reduction of all types of arms and arms preparation, otherwise it is spurious. This means that it must include the progressive limitation of secrecy in all types of research, including even military research. The institution of these radical changes requires a rapid reorientation of people everywhere and especially of their political leaders.

This about-face can be successful only if strong, positive efforts are made to engage the people of both sides in joint enterprises: scientific, cultural, commercial, and industrial. They must be enlisted, more especially, in the urgent job of giving substantial help to the technically less advanced peoples. This will be at the cost of real but only temporary sacrifices on their own part. Finally, in order that this reorientation be put on a solid basis, education everywhere must be thoroughly revised, to bring to all peoples the scientific world view, the unifying view whose foundations are continually open to inspection, criticism, testing, and revision. This common view will show all peoples the greatness of the world, and the sublime possibilities open to all men when they use their minds rationally and freely, and when they give their hearts full scope.

The promotion of this planet-wide cooperation calls to an increasing degree for world law, which needs eventually to be backed by adequate international force. Under such law, however, each country must be allowed to develop freely those ways of life that it finds invaluable. For the fruits of cultural evolution must not be lost, and these diversities will enrich the whole. Finally, then, the aggressive impulses of men will be channelled into constructive rivalries, and into the mighty struggle to subjugate nature for the common good.

Human Values in Relation to Evolution

BY something of value we mean, of course, something desirable, to be striven for, and from the attainment of which we presumably derive satisfaction. The value of a thing, in other words, is its property of being desirable. If it is undesirable, its value may be said to be negative. By these definitions, all animals that can pursue conscious purposes recognize values in given situations, values partly built in as such and partly channelized and modified by experiences, and these values are of different degrees, so that they could be represented on a scale. Probably only man, however, of animals on this earth, has reached the stage of pondering over values, and singling them out, as it were, by speech symbols. And so only he has attempted classifications for grouping things of related value together and has arranged these groups in hierarchies, with higher rank accorded to those expressing certain general psychological principles, such as heroism or truth. Nevetheless, even the non-human animal, like the young child, often acts according to much the same principles but without abstracting them.

When the matter is regarded in this way, it becomes evident that values are as legitimate a subject of scientific investigation as any other phenomena pertaining to living things. In fact, it is imperative that they be investigated by scientific methods if we would seek greater clarity and assurance regarding our own aims and if we would order and shape these aims in such wise as to attain, in human living, a more harmonious interplay of strivings and of activities, and a deeper sense of fulfillment. The pursuit of this project, however, has unfortunately been hampered by the prevalence of two tenets that are incompatible with it. The earlier, historically, of these tenets holds that values for man are properly determined only by some higher authority, external to him. The second tenet, often thought of as an alternative to the first and yet related and reconcilable with it, holds that, for each individual, questions of value are purely his own private matter, shared only with those who already agree with him, as in the case of some schools of artists, and that these questions cannot profitably be argued about. My discussion is not directed to those who are irrevocably committed to either of these views.

Biological Significance of Values

There are no grounds in evolution theory for regarding the likes and dislikes of nonhuman animals, their emotional reactions and drives, or their resultant desires and values, as being determined by processes any different in essentials from the evolutionary mechanisms whereby their other characteristics were laid down. The same principles of blind mutation, genetic recombination, and natural selection as have been inferred to underlie the evolution of other physiological as well as morphological peculiarities are equally applicable to the genesis and establishment of the so-called affective traits. It is therefore not surprising that the present-day emotional and affective equipment of any species provides the basis for an intricately interrelated web of responses that are all nicely adjusted to the service of the same major end—species survival and expansion—even as is true of all other bodily functions.

This by no means implies, however, that each species is constructed in the best possible way for leading its own kind of life in its own kind of niche nor even that there is any one best way for doing so. For of course the products of evolution, although displaying within themselves amazing coordination, are patchworks thrown together bit by bit without long-range foresight. Time and again they have had to be remodeled opportunistically, in one small way after another, as new emergencies or opportunities have arisen. Undoubtedly, if the organisms could have been made from blueprints designed from the first to enable them to lead their present types of lives under their present circumstances, diverse disharmonies, indirections, and complexities that are now incorporated solidly in their structures could have been avoided, with considerable gain in effectiveness.

In the case of a species which, like man, has changed its mode of life rather recently, in terms of geological time, the remodelings are still relatively crude, conspicuous, and on occasion troublesome. Thus more friction, both inner and outer, would be generated in such species than in the far more numerous ones that had kept running in pretty much their old established ways. All this may be expected to hold in the realm of "behavior," including that of desires and values, just as it does in the other modes of functioning of the organism. However, in higher forms and more especially in man, the plasticity of their behavior would tend to conceal defects in its bases.

A consideration of the values that, in any species, are attached to diverse experiences, in situations involving relatively little modification by conditioning or association, necessarily involves some understanding of the entire complex system of drives, response potentialities, thought processes, and sensations, along with the affective mechanisms, including emotions, attached to these activities, in a creature of the given type. Science is at present woefully far from a detailed analysis of these matters. It has been generally conceded that in insects there are indeed a multitude of highly particularized and often intricate patterns of behavior that are to a very large extent genetically determined. Moreover, fish and especially birds are hardly behind insects in this respect, even though their reactions are more open to modification by experience. In the case of mammals, it has been widely held that the genetic behavior patterns are less specific, both in regard to their instigating stimuli and to their methods of response, and that in man they have virtually disappeared to make way for reactions determined almost completely by experience.

Closer scrutiny shows, however, that despite the enormous increase in the effectiveness and complexity of associations in man, his nature provides him with a great number of unlearned predilections and aversions. These range from those on a purely sensory level, such as a liking for sweets, diverse savories and fragrances, feelings of softness and smoothness, warmth, harmonies, color combinations, varied kinesthetic and other bodily feelings, and so forth, and a dislike for their opposites, to those on a perceptual and perhaps even a conceptual level.

Genetically Based Differences in Values

Of course perceptions and conceptions are gained only by learning. Yet, curiously enough, individuals display marked inheritable differences in regard to the strength of feelings attached to certain concepts, such as the feelings that have been denoted by the terms *egotism* and *sociability*. For example, it requires considerable experience, analysis, and association of sensations and perceptions before one arrives at the concept of one's self and of other selves, yet it seems much as if there were a mechanism lying in wait to attach positive affective values to these concepts when or shortly after they appear. Perhaps this effect represents only the combination of the values already attached to the components of the concept, plus those later added by association. However that may be, self-esteem may become developed much more strongly in one person than in another even though the experiences of the two seem to have been essentially similar, and even though one may seem to have had no more attachment for the separate elements out of which the ego conception was built than the other person has had.

We are far from understanding the built-in mechanisms by which such results come about, but their efficacy is attested in varied ways, both in man and in other more-or-less intelligent animals. Among dogs, for instance, such differences in temperament, as we call them, have long been observed to be associated with particular breeds, even when the young are taken from their parents so early that the behavioral traits in question could not have been acquired purely by experience. At the same time, an acknowledgment of this situation should not lead us to minimize the enormous role of experience, especially in man, in channelizing and shaping the modes of expression of these feelings, and in determining what situations call them forth.

It would be presumptuous and fatuous, in our present ignorance, to attempt here to list or classify the numberless inherent affective tendencies—that is, natural predilections and aversions—observable in mammals, the great majority of which also appear to be present in some degree in man. It should be noted, however, that in those classifications that include only a few general categories, such as fear, anger, love, and curiosity, each item really represents a complex group of natural tendencies that are in some manner related. For instance, under fear would come fear of heights, which recent experiments with nonhuman mammals have shown to be expressed in them in the absence of previous experience, and which undoubtedly evokes somewhat different emotions as well as responses from those found in other innate fears, such as fear of loud deep tones having the quality of thunder or a growl. Again, sexual love is itself very complex in its inherent basis, and although it has elements in common with or similar to those in the parental, filial, and communal forms of attachment, it is misleading to represent it as differentiated from the others, or them from one another, solely as a result of experience. Moreover, in addition to the inherent affective tendencies included under a few general headings such as those mentioned, there are many that do not fit under any of them, such as those that participate in our multiform experiences of beauty and revulsion, of exhilaration and depression, and so on.

Finally, it should be realized that, contrary to the assumption of some authors, the fact that a given psychological trait appears later than at birth by no means affords evidence that it is purely an acquirement gained by experience. Inherent affective traits, like all other characteristics of organisms, tend not to attain manifestation before a given stage, peculiar to that trait. However, because of the enormous influence that is also exerted by experience in the manifestation of traits, the disentanglement of the processes at work in the case of a given trait often becomes very difficult.

Despite a large amount of agreement between different species of mammals in regard to their constellation of affective tendencies and the way in which these attain expression, there are of course considerable differences between species and even subspecies in the relative strengths of these tendencies, in their mode of expression, and more especially, in the stimuli that evoke them, in adaptation to the differences in the organisms' modes of life and capabilities. To this extent, each type of creature pursues somewhat different values from the other creatures and different immediate aims. It pursues some of these for their own sake, with due compromise amongst them. These subjective ends are so adjusted that for each species and subspecies of animal, although of course not for each deviant individual, this compromise tends to work in the direction of the ultimate objective end of the species: its maintenance and spread. But the creature does not recognize this end as such, nor does it recognize even the more limited end of "self preservation" that is popularly attributed to it.

Genetic Foundation of Altruistic Values

Of course, the subjective ends of a creature may be far from purely selfish. First Darwin, and later Prince Peter Kropotkin, pointed out that natural selection is not just a tooth-and-claw struggle. Insofar as cooperative behavior furthers survival, natural selection has promoted emotional bonds and a spirit of service among members of the same family and the same community. However, this form of selection is limited to situations in which an individual, in helping others, by these very acts somehow assists in the survival of its own genes, or of the same or similarly acting genes in the other individuals—that is, genes of the type that led to these cooperative acts. Obviously the nurturing and protection of the young belongs in this category, since the individual is thereby fostering genes derived from itself. However, many ways of life put a premium on early dispersal of the young and in these cases maternal care is kept minimal. Similarly, "rugged individualism" rather than altruism among adults would tend to win out in the numerous species in which dispersal of adults is advantageous, or in which means of mutual aid, in the sense defined above, are hard to come by.

Fortunately for the mammals, their mammary glands, apparently originally derived from the more generalized sebaceous glands that serviced their fur, provided a ready means of nourishing the young and thus laid the physical basis upon which a high development of all the maternal impulses, including especially maternal love, became practicable. It is evident that in many cases some extension, with suitable modifications, of these feelings and modes of response to the relations between father and young, father and mother, and finally among all immediate relatives, must have proved helpful to the propagation of the genes concerned in these developments. Under certain circumstances, including those of ground-dwelling primates, which are relatively defenseless individually, there proved to be comparative safety from predators in numbers, and also other advantages, as in hunting, when it emerged, in sharing fire, and in sharing services and abilities. From then on, genetic variations leading to associations among families, and to a partial extension of the affective family ties to the other members of the community, were rewarded by enhanced survival and multiplication. All this meant a step-by-small-step development of the genetic bases of social feelings and behavior—that is, of the complex sometimes comprised under the term *brotherly love*.

It is important to note, however, a self-limiting feature of the genetic process underlying this development. So long as the communities are small, variants having genes that predispose them to helping the members of other families will be able appreciably to foster the survival and spread of their community as a whole. This community, being small and interrelated, will usually contain a higher proportion of genes of the type in question that lead to this cooperative action, than another community in which there is less cooperation and which therefore is less successful. Thus the cooperative individual tends to further the survival and spread of genes like his own even though he himself is sacrificed. However, as this very process and the cultural evolution attending it cause these groups to grow larger, the power of this strictly *inter-group* selection diminishes. Perhaps the resultant slowing down of natural selection for social traits affords a partial interpretation of the relatively low development of the broader social impulses in present-day man.

As has often been pointed out, social insects which have only a few reproducing individuals per colony do not suffer from this self-limitation of natural selection for social traits when the number of sterile workers per colony increases. These workers represent only somatic extensions of a very restricted colonial collection of genes. Thus the gene composition has the high inter-colony variability of small numbers, and a high degree of self-sacrificing behavior on behalf of the colony reflects and tends to perpetuate and spread the kind of genotype that has produced this behavior. In this way these insects have been able to attain a far higher degree of *genetically based* socialization than has man. Nevertheless, their potentialities are of course far more limited, by their inordinately lower intelligence and capacity to learn, their inability to accumulate culture, and their inability to attain a realization of what the fuss is all about.

It has sometimes been held, as for instance by some schools of Communists, that the development of social feelings in man is merely an expression of enlightened self-interest associated with his intellectual, technical, and cultural development, without any important basis in his inherent affective or emotional peculiarities. A genetically asocial but intelligent man, if already living in a group where social feelings and practices prevailed, would doubtless try to camouflage himself, for his own good, into a seeming replica of his fellows. But this would not prevent the lone wolf in sheep's clothing from practicing self-aggrandizement at the expense of others wherever he judged he could get away with it. A community of such individuals could not hang together indefinitely. Still less would they *voluntarily* organize themselves into a socially behaving group, devoid of genetic sheep. The advantage of mutual aid to a group thus depends in the main upon its individuals practicing it even where they themselves do not reap any profit other than the intrinsic reward of such acts to those for whom such behavior is a primary end. Once such a feeling is present, however, its expression can be reinforced by various accessory motivations, such as pride in following out the rules of the community.

Cultural Elaboration of Values

We cannot enter here into the manner in which, in increasingly intelligent animals, the pattern of behavior, and of affective tendencies, is more and more modified by experience. Still less can we here review the truly revolutionary innovation in method whereby, in man, experience becomes transferred, symbolized in speech and otherwise, accumulated, codified, and sanctified. It is important, however, to realize that it is a function of these processes of modification, whether by individual or by group experience, to extend the inherent affective tendencies to situations which the experiences have shown to be somewhat connected with the ones that originally aroused the given feelings. Thus, for each individual or, where experience is transferred, for each group a pattern of feelings and responses becomes built up that has, so to speak, been recut so as better to meet the set of circumstances peculiar to that individual or group. Yet the old feelings are in large measure still there, and in many cases they are still evoked by the original stimuli, as well as by the associated ones.

It is true that the associations form extended chains, or rather webs, yet the motivations that trigger the decisions to act or not to act are in the last instance based in the ingrained affective components that are directly or indirectly attached to elements in any given web, and that have now become more or less suffused throughout that web. It should also be noted that, within any web, not all affective components need work in the same direction. Thus the attainment of the final decision may entail considerable inner friction.

The transfer and accumulation of experience—that is, cultural evolution—was of course possible only because of the intelligence, social disposition, and manipulatory, vocal, and other special proclivities of genetic origin that had laid a basis for it. For a very long time, while culture was being accumulated, this genetic basis also must have been strengthened by natural selection, and the cultural process must have been facilitated thereby. More recently there has probably been a slowing down of the genetic advance, caused by the formation of fewer, larger groups and by the greater efficiency of mutual aid in helping the less fit.

Another factor that must have facilitated cultural evolution in the past is a kind of nongenetic natural selection operating between different groups, and between portions of a group, so as to favor more the continuance and spread of those whose cultures (as distinguished from their genes) were more conducive to their own survival and increase. This process, emphasized in its intergroup aspects by S. J. Holmes, tended to promote the more functional cultures and to curb the biologically unsounder ones that went off on cultural tangents by a kind of inner inertia. Undoubtedly this natural selection of cultural features also went on within the group, to favor the survival and multiplication of those whose traditions of value and conduct were more functional, even as, within our own society, the Shakers, whose religion forbids them to reproduce, have virtually died out. The situation is of course different with a celibate priestly caste since this is automatically replenished from the rest, although it can of course result in the gradual depletion of superior genetic material in the group as a whole.

Nevertheless, the development of culture proceeds primarily through its own operations. Although the core of these is the joint accumulation of experience, most of that experience until modern times has consisted of purely empirical observations and familiarity with rule-of-thumb operations, with little understanding of the nature of the matters involved. The artificial interpretations, almost always animistic, that were gradually fabricated regarding the nature of the world and man, and the associated rituals, witch-doctoring, and institutionalization of superstitution, did however fill the important general function of reinforcing the social solidarity of the group and promoting their cooperative behavior. They also played important technical roles in healing, in advantageously regulating hunting, planting, and so on, and in laying down rules of conduct that were made to seem dependent on the interpretations themselves. Our modern scientific knowledge makes the intellectual fallacies and the technical imperfections of these primitive systems so glaring that we tend to forget the positive role they played until we see a primitive people lose all interest in life and disintegrate when modern civilization deprives them of their Rock of Ages.

The religious and ethical systems of nonscientific peoples expressed the values that they overtly recognized, but in doing so they gave opportunity for the satisfaction of the peoples' actual affective needs and resulted in behavior conducive to the survival and extension of the group. Despite marked differences among these systems, in viewpoint and methods, remarkable similarities are also displayed, especially among the more successful ones, reflecting the underlying likenesses in human psychological and physical needs, combined with plasticity in adjustment to these needs.

All the more successful social systems emphasized by precept and training the importance of service to others, not only in the family (which of course had priority) but also in the group as a whole, and, later, more especially, to the leaders. Moreover, those specific attitudes and practices were fostered that tended to make that service more effective: practices such as veracity, integrity, self-control, industry, and courage. In addition, devices were used for arousing ecstatic emotional experiences that deepened and intensified the individual's sense of forming a part of a greater family dominated by a greater father, who provided greater rewards and more frightful punishments. These experiences also enhanced the individual's sense of privilege and achievement in participating in the activities of this supernal fellowship.

Revision of Values under Civilization

Until some 2,500 years ago community loyalty was usually accompanied by a then healthy suspicion and even hostility toward other communities, especially those with different cultures, and often by a zeal in striving against them that matched and nourished the intragroup cohesion. But with the rise of the great empires that embraced many previously separate peoples, doctrines of brotherhood among all mankind began to gain increasing acceptance. Along with this there was a growing adherence to abstract conceptions that were supposed to embody universally valid higher values, such as Plato's "Goodness, Beauty, and Truth." However, these terms were seldom defined concretely enough, in view of the innumerable interpretations which could be given them, to provide unequivocal solutions to problems of actual living on the part of either the individual or the group. Thus, Plato regarded slavery as fundamental for his system of Goodness, Beauty, and Truth.

It is high time for modern man, everywhere, again to revise his concepts of values, in accord with the utterly new view that science, and especially evolutionary science, has given him of the nature of the world and of his actual and potential relations to it. We must admit that it is much too early for detailed formulations of the place to be accorded to the diverse major and minor values that flow out of his numerous inherent affective tendencies and out of the possibilities of interconnecting them, modifying their modes of instigation and expression, enhancing some and subordinating others. For this purpose we have as yet far too little knowledge of just what these tendencies are, and of their genetic and acquired variations, in neurological, psychological, and sociological terms. Yet we can already discern clearly certain major features that lead to important general conclusions.

Man is not made with any one inherent affective tendency or drive and associated value that can normally provide a clearly overriding aim for all his existence. It is true that *deprivation* of the opportunity to pursue a given drive may cause its value to become overriding. But be this drive sexual love, love of offspring, of his fellows, or of himself, be it joy in domination, or in subordination, be it love of variation or of the familiar, or of good eating, or anything else, man is normally a great bundle of natural and acquired wishes and values, interrelated in a pattern that is partly individual but very largely a product of his group's cultural evolution, including the contributions made by great and little thinkers of his past. This entire complex has in its general lines been framed so that the working of this system tends on the whole to promote the survival and multiplication both of his family and of his larger group. Thus his values tend to be realized more fully when this objective end, even if not understood, is promoted by him.

It is natural that this should be the case, since in the first place natural selection has worked to provide him with predilections and aversions that under primitive conditions would combine to lead to this result. In the second place, his culture has on the whole worked to modify the directions of these wishes and struggles so as to lead to this result under the conditions of that culture. In recent times, however, human understanding and the conditions of human living and association have been changing so fast that our systems of values and ethics, education, and social relations, have fallen behind and are not well enough adjusted to our present needs and knowledge.

What kind of reorientation is needed? Obviously, one in which our motivations are reshaped in accordance with our modern knowledge of the world and ourselves.

Among the deeply and complexly rooted inclinations of men is their readiness to accept the challenge of any situation confronting them, if it affords a hope, through struggle, of winning through to greater life for themselves and for those with whom they identify themselves. With the shrinkage of the world, and the suicide that is being committed by war, if it does not murder us first, each man must ever more strongly identify himself with humanity in general. The visions that he has obtained of the unimaginable progression already accomplished in past evolution, of the unprecedented powers which he himself has now gained through science, and of the fathomless reaches to which man may go, in terms of greater life, by the rational use of these powers in behalf of himself and posterity, afford an overall directive for his efforts that is in accord with the objective end of the species—namely, its survival and extension—and also with most of his own more immediate subjectively based values. Enough can thereby be gained for the individual, in enhanced richness and harmony of life, to recompense him on a personal basis to a degree unparalleled in the past, especially if we will take advantage of already existing psychology and psychiatry. At the same time, he can attain a sense of participation in a joint endeavor far greater than his own that is more solidly based and more buoying to his spirits than that gained by obedience to a tenuous superior power.

As between individual and social values (or welfare) there is no ultimate contradiction for beings who have the natural feelings that lead to mutual aid. However, until we have gone further in the development of these feelings, through both training and genetics, there will be many situations calling for difficult compromises on our part. It is important in such connections to remember that those men who have had better opportunities for all-round development and zest in life, as a result of their personal experiences and activities, and who have a significant voice in the determination of their own destinies, can also give better service to the community.

Higher Values

It may here be objected that we have not yet given an opinion as to what should be considered the higher subjective values that are most to be striven for. If objectively we take survival and extension of the species to be the end, as set by evolution, then we find that in our own line of descent the two groups of psychological characteristics that have been the most important in putting us into our dominant position were those making for intelligence and those making for cooperative behavior. Moreover, despite the revolutionary changes wrought by our culture since its creation of modern science, it is evident that these two functions, especially when enhanced through cultural measures and, I believe, eventually through genetic measures also, will *continue* to provide the most important means of meeting the evolutionary test of survival and extension, while at the same time they will render us ever more capable of filling not merely our physical but also our psychological needs—that is, of achieving and advancing our values.

In accordance with this thesis, we may place in the foremost positions among psychological needs, and we may accord the highest value to, for one thing, the gratification of curiosity—that is, the pursuit of truth for its own sake, by methods of the most effective kind (the kind used by science)—and, for another thing that is no less important, the fulfillment of love in its varied aspects. Among other values, the cultivation of which is also highly important, we may mention here as a few examples, largely overlapping with one another and with the two already given, the zest for making one's own decisions (that is, for the exercise of freedom), for achievement, creativity, variety, and adventure, and the appreciation of nature, art, and artifice. All of these overlapping values can be harmonized with one another, and the seeking of them will play major functional roles in our objective as well as subjective progression.

It has been rightly said that biological evolution is multidirectional and cruel and that the vast majority of lines of descent end in pitiful anticlimaxes. Yet it is also true that intelligence and cooperation, culminating in ourselves, have not merely constituted one of the lines of nature's development, but *the* one which, through its increasing control of the rest of nature in its own interests, has become by far the most prominent. But this development is only beginning. It is our business to take it as much farther as we can, in the creativity of our coordinated, voluntary efforts. Through the unprecedented human faculty of long-range foresight, jointly serviced and exercised by us, we can, in securing and advancing our position, increasingly avoid the missteps of blind nature, circumvent its cruelties, reform our own natures, and enhance our own values.

The foregoing conclusions represent, I believe, an outgrowth of the thesis of modern humanism, as well as of the study of evolution, that the primary job for man is to promote his own welfare and advancement, both that of his members considered individually and that of the all-inclusive group, in due awareness of the world as it is, and on the basis of a naturalistic, scientific ethics.

Social Biology and Population Improvement

IN response to a request from Science Service, of Washington, D.C., for a reply to the question, "How could the world's population be improved most effectively genetically?" which was addressed to a number of scientific workers, the subjoined statement was prepared, and signed by those whose names appear at the end.

The question, "How could the world's population be improved most effectively genetically?," raises far broader problems than the purely biological ones, problems which the biologist unavoidably encounters as soon as he tries to get the principles of his own special field put into practice. For the effective genetic improvement of mankind is dependent upon major changes in social conditions and correlative changes in human attitudes. In the first place, there can be no valid basis for estimating and comparing the intrinsic worth of different individuals, without economic and social conditions which provide approximately equal opportunities for all members of society instead of stratifying them from birth into classes with widely different privileges.

The second major hindrance to genetic improvement lies in the economic and political conditions which foster antagonism between different peoples, nations, and "races." The removal of race prejudices and of the unscientific doctrine that good or bad genes are the monopoly of particular peoples or of persons with features of a given kind will not be possible, however, before the conditions which make for war and economic exploitation have been eliminated. This requires some effective sort of federation of the whole world, based on the common interests of all its peoples.

Thirdly, it cannot be expected that the raising of children will be influenced actively by considerations of the worth of future generations unless parents in general have a very considerable economic security and unless they are extended such adequate economic, medical, educational and other aids in the bearing and rearing of each additional child that the having of more children does not overburden either of them. As the woman is more especially affected by child-bearing and rearing, she must be given special protection to ensure that her reproductive duties do not interfere too greatly with her opportunities to participate in the life and work of the community at large. These objects cannot be achieved unless there is an organization of production primarily for the benefit of consumer and worker, unless the conditions of employment are adapted to the needs of parents and especially of mothers, and unless dwellings, towns, and community services generally are reshaped with the good of children as one of their main objectives.

A fourth prerequisite for effective genetic improvement is the legalization, the universal dissemination, and the further development through scientific investigation, of ever more efficacious means of birth control, both negative and positive, that can be put into effect at all stages of the reproductive process—as by voluntary temporary or permanent sterilization, contraception, abortion (as a third line of defense), control of fertility and of the sexual cycle, artificial insemination, etc. Along with all this the development of social consciousness and responsibility in regard to the production of children is required, and this cannot be expected to be operative unless the above-mentioned economic and social conditions for its fulfillment are present, and unless the superstitious attitude towards sex and reproduction now prevalent has been replaced by a scientific and social attitude. This will result in its being regarded as an honor and privilege, if not a duty, for a mother, married or unmarried, or for a couple, to have the best children possible, both in respect of their upbringing and of their genetic endowment, even where the latter would mean an artificial —though always voluntary—control over the process of parenthood.

Before people in general, or the State which is supposed to represent them, can be relied upon to adopt rational policies for the guidance of their reproduction, there will have to be, fifthly, a far wider spread of knowledge of biological principles and of recognition of the truth that both environment and heredity constitute dominating and inescapable complementary factors in human well-being, but factors both of which are under the potential control of man and admit of unlimited but interdependent progress. Betterment of environmental conditions enhances the opportunities for genetic betterment in the ways above indicated. But it must also be understood that the effect of the bettered environment is not a direct one on the germ cells and that the Lamarckian doctrine is fallacious, according to which the children of parents who have had better opportunities for physical and mental development inherit these improvements biologically, and according to which, in consequence, the dominant classes and peoples would have become genetically superior to the underprivileged ones. The intrinsic (genetic) characteristics of any generation can be better than those of the preceding generation only as a result of some kind of *selection,* that is, by those persons of the preceding generation who had a better genetic equipment having produced more offspring, on the whole, than the rest, either through conscious choice, or as an automatic result of the way in which they lived. Under modern civilized conditions such selection is far less likely to be automatic than under primitive conditions, hence some kind of conscious guidance of selection is called for. To make this possible, however, the population must first appreciate the force of the above principles, and the social value which a wisely guided selection would have.

Sixthly, conscious selection requires, in addition, an agreed direction or directions for selection to take, and these directions cannot be social ones, that is, for the good of mankind at large, unless social motives predominate in society. This in turn implies its socialized organization. The most important genetic objectives, from a social point of view, are the improvement of those genetic characteristics which make (a) for health, (b) for the complex called intelligence, and (c) for those temperamental qualities which favor fellow-feeling and social behavior rather than those (today most esteemed by many) which make for personal "success," as success is usually understood at present.

A more widespread understanding of biological principles will bring with it the realization that much more than the prevention of genetic deterioration is to be sought for, and that the raising of the level of the average of the population nearly to that of the highest now existing in isolated individuals, in regard to physical well-being, intelligence, and temperamental qualities, is an achievement that would—so far as purely genetic considerations are concerned—be physically possible within a comparatively small number of generations. Thus everyone might look upon "genius," combined of course with stability, as his birthright. As the course of evolution shows, this would represent no final stage at all, but only an earnest of still further progress in the future.

The effectiveness of such progress, however, would demand increasingly extensive and intensive research in human genetics and in the numerous fields of investigation correlated therewith. This would involve the cooperation of specialists in various branches of medicine, psychology, chemistry and, not least, the social sciences, with the improvement of the inner constitution of man himself as their central theme. The organization of the human body is marvelously intricate, and the study of its genetics is beset with special difficulties which require the prosecution of research in this field to be on a much vaster scale, as well as more exact and analytical, than hitherto contemplated. This can, however, come about when men's minds are turned from war and hate and the struggle for the elementary means of subsistence to larger aims, pursued in common.

The day when economic reconstruction will reach the stage where such human forces will be released is not yet, but it is the task of this generation to prepare for it, and all steps along the way will represent a gain, not only for the possibilities of the ultimate genetic improvement of man, to a degree seldom dreamed of hitherto, but at the same time, more directly, for human mastery over those more immediate evils which are so threatening our modern civilization.

Among the signatories of Muller's statement were:

F. A. E. Crew	R. A. Emerson
C. D. Darlington	C. Gordon
J. B. S. Haldane	J. Hammond
S. C. Harland	C. L. Huskins
L. T. Hogben	P. C. Koller
J. S. Huxley	W. Landauer
J. Needham	H. H. Plough
G. P. Child	B. Price
P. R. David	J. Schultz
G. Dahlberg	A. G. Steinberg
Th. Dobzhansky	C. H. Waddington

What Genetic Course Will Man Steer?

OF course we—that is, humanity—will take our biological evolution into our own hands and try to steer its direction, provided that we, humanity, survive our present crises. Have we not eventually utilized, for better or worse, all materials, processes, and powers over which we could gain some mastery? And are there not means already by which we can influence our heredity, and other means that we are likely to gain? Some detractors may call the use of them tampering, tinkering, or even blasphemy. But this attitude is like that of recalcitrants who still hold out against whatever foods, remedies, or measures they consider "unnatural" (e.g., the Amish who even today consider higher education to be a perversion of the mind). Any of these things can be disastrously misused, or they can be used to great advantage.

It is life's essence to utilize, wherever possible, more and more effective means of servicing itself. That implies doing things in ways that were previously unnatural. Moreover, the rise of man to ascendancy over all other forms of life has resulted from his having been so unusually successful in this very respect. This has been true in both his genetic and his cultural evolution. Indeed, man is the first organism for whom any culture except a trace has become natural. That is, increasingly, the ways and products of culture, the so-called artificial, have been naturalized by man. All this, however, is far from saying that every new or artificial way or product is a sound one. The so-called "primrose path" is readily found, and it is wider than the so-called "straight and narrow one." That is why there must be a most far-sighted, judicious, and beneficent steering by man in general of both his genetic course and his cultural one. These must in fact be intermeshed and complementary to one another.

Life the Lucky

For our present purposes, we may define genetic advances as the gaining of abilities for making use of the environment more effectively, and for withstanding or even making use of circumstances that earlier would have been useless or hostile. By this measure, the totality of living things—that is, all species taken together—has certainly advanced enormously through the ages. For it has increasingly extended life's domain, increased its resources, and made it more secure. Moreover, certain lines of descent, most notably the one leading to ourselves, have ultimately advanced the most by these criteria, or, as we say, they have become the highest. They, and especially we, are the ones that can overcome the greatest difficulties, and the most adverse ones. And so, even though we certainly have our worries, these are largely of our own making. And we, the self-styled heirs of all the ages, constitute the very luckiest, the most improbably lucky, combination of trials of the whole lot.

The luck that allowed any line to advance genetically was of course based on the Darwinian natural selection of mutant types and of combinations of them. Since the kind of mutation occurring cannot be influenced by the effect it will have, and since there can be ever so many more ways of harming than of improving any mechanism, vastly more mutations, and combinations of them, proved to be failures rather than successes in their influence on "survival." Here "survival" must be understood to include multiplication, and so it would be better to say "genetic survival," "net multiplication," or simply "fitness." It is the *multiplication* of the successful mutants that plays the key role in evolution, for it alone allows for additional successful steps. To permit room and resources for this multiplication there must of course be correlative reduction in numbers, or extinction of some less successful types, when the new ones that succeed all go into virgin territory, or somehow make enough extra resources available for the others too.

Thus it is clear that "natural selection," or "the survival of the fittest," really means the *differentially* high net-multiplication, in some environment, of certain rare mutant types which we call "the fitter" or "fittest." This process has been made possible, even inevitable, by the gene-material's unique properties of replicating, as such, even its mutant forms, and accumulating them to an unlimited degree within one chromosome-set. Moreover, despite the rarity of serviceable mutations and the minuteness of the effects that individual serviceable mutations usually produce, the speed of evolution has been greater than these circumstances would seem to allow, for the getting together of the mutant genes has been greatly aided by sexual recombination. The eons of time over which they have accumulated has allowed the formation of such marvelous organizations, having such consummate integration, as protoplasm, and eventually man himself.

Our Forebears' Run of Luck

Let us review briefly some clues to man's concatenation of luck having been so much greater than that of other organisms by focussing upon his ancestors of the last hundred million years, the primates. A long succession of events had already made the mammals the most advanced animals. Of the primates, remains of the most primitive known group, the prosimians, which today includes the lemurs, have been found in strata that also contain remains of dinosaurs. Prosimians must early have gained (if not inherited) such physical advantages for active life in trees as opposable first digits, improved vision, equipment for a somewhat omnivorous diet, and uniparity. But they were soon pushed into the background and thus hampered in advancing further by their more successful offshoots, the simians, that is, monkeys and apes. Thus, they failed to gain the simians' greater maneuverability, curiosity, and general intelligence.

However, George Joyce's recent studies in the field show that lemurs do have a highly developed maternal solicitude—as demanded of animals that must take constant heed and care of their single young. There is a similarly high level of intra-group cooperation at all ages, which is a kind of extension of maternal love and empathy to companions in a permanent group whose members include individuals of both sexes. There is also play between adults. And so, what might be called social intelligence was enabled to flourish. It is important to note that this faculty probably preceded the kinds of intelligence concerned with inanimate objects and with other types of organisms. Moreover, there goes with this intelligence much learning, from elders and from playmates, of what behavior to adopt in given situations, what foods to seek, what to reject, and so on. This receptiveness to the attitudes and acts of others affords an important part of the basis needed for our own cultural evolution.

The bodily and psychological advances made by the monkeys and apes gave a further basis for the advances afterward made by the apes' protohuman off-shoot, which split off from the other apes some twenty million years ago. However, they have been too much discussed to be closely reviewed here. So have the factors that probably led to these traits having been favored by natural selection. Suffice it here to call attention, on the so-called physical side, to the constant view forward, with its opening up of wider opportunities, permitted by the apes' arm-mobility and consequent arm-swung mode of progression and, derived from the latter, their semierect posture even on the ground.

These traits put even more of a premium on broad awareness, versatility, and love of variety, hence too on curiosity concerning objects—both inanimate and animate—and general intelligence. The latter includes a higher ability to transfer lessons learned in a given field to another one, and to solve problems. This in turn allowed, at least in the chimpanzee, the making of very simple tools and some hunting of game.

Meanwhile, social intelligence, affection between companions, and cooperation had also developed. For the little groups had continued to be fairly permanent, and to include individuals of all ages and could therefore profit by emphasis on these social traits. Moreover, the division into many small social groups must have promoted natural selection for the genetic bases of social intelligence and of social traits in general. This is because genes that tend to extend maternal and brotherly feelings to other members of the closely related little group result also in mutual aid. By thus helping the group's survival, these genes actually foster their own survival even when they lead to self-sacrifice, since others of the tiny band tend to have the same genes. By the greater growth, followed by the resplitting of the more social little groups, the genetic groundwork of cooperation was increasingly strengthened in the species.

Man Improves on Luck

In these ways, the genetic structure must have been laid down for a line of descent which, branching off from that of other apes some twenty million years ago, could by virtue of both its bodily and mental traits get along increasingly well on the ground by defending itself better from predators and in various other ways. By some two million years ago, its members had already become fully erect and much like ourselves in form, except for their little more than ape-size brains and rather large jaws. Since their lairs contain abundant broken bones of fair-size game, as well as rough-hewn tools, they must not only have evolved much more initiative, including aggressiveness, than apes but also, and most important, they must already have accumulated a substantial amount of extragenically transmitted experience. In other words, cultural evolution, a process so nearly unique in the human line, had begun in earnest.

Like the evolution of the genetic constitution, that of culture requires the arising, the transmitting, and selecting of innovations. But since the cultural innovations are in thought and behavior, their transmission is by some form of imitation, not heredity, even though genes must afford the abilities for these processes. Of course this form of transmission allows a much more rapid spreading than that through differential multiplication. But in the earlier stages the acceptable innovations, like acceptable mutations, tended to be rare and apparently insignificant, for they arose in a rather haphazard way, in which foresight was extremely limited. Therefore, like mutations, they had to be selected after their origin, according to their helpfulness to the individual and his group.

But as culture very slowly accumulated by these means, the rate of development gradually increased. This was not only because culture itself affords means of producing more culture. It was also because in those times the use of culture afforded more opportunity for the natural selection of the genetic traits which allowed that culture to be utilized in ways serviceable to the particular user and the user's little group. Hence culture must have reacted, via natural selection, to enhance the genetic foundations of cooperativeness, initiative, general intelligence, and such more special faculties of mind and body as facilitated the use or the accumulation of culture itself, or caused adaptations to the conditions arising from culture. Foremost among these special genetic faculties was that of communication, especially speech, which in turn depends on a complex of unusual propensities. All these genetic advances allowed, in their turn, faster cultural evolution.

Thus there was for a very long time a "positive feedback," of reciprocal nature, between evolution of the genetic and the cultural types. So on the genetic side, it is not surprising that during the past two million years brain size (which is but one of many factors in mental ability) underwent a most drastic increase—about a tripling. And, as we all know, cultural evolution had, by some ten or twelve thousand years ago in some regions, reached such a rate that it resulted in the successive breakthroughs (so close together in terms of evolutionary time) of food-growing, then town-dwelling, and now our modern sciences and technologies.

During the course of these developments human foresight as well as hindsight became enhanced. Hence the initiation of cultural innovations gradually, and with the scientific technological breakthrough very rapidly, became less haphazard, unlike that of mutations. They could increasingly be preselected to advantage, more reliably and rapidly post-tested, and their transmission became faster and more diverse. Larger steps then became more feasible, and even necessary, and they were and are being taken. Imagination and foresight are supplanting luck through trial and error, although they can never do so completely.

Meanwhile, with the improved techniques of production and transportation of even the prescientific stages of urban life, the groups of people, originally so tiny, grew in density and area and merged increasingly, or were forced to do so. Thus great states emerged in diverse regions, and religions sprang up that emphasized the brotherhood of all men. Today, of course, it has become both possible and necessary, if civilization is not to founder, for all groups to federate together, within a surprisingly few scores of years, into one great community, all of whose people share comparably in the fruits of modern technology, in the modern scientific world-view, and in human dignity and opportunity. The convergence of attitudes and cultures has recently been admirably set forth in a statement, *Education and the Spirit of Science,* issued by the Educational Policies Commission of the National Education Association.

In this community, no place can be left for biases against races or social classes—or else! Hence, racial amalgamation will gradually and voluntarily but inevitably ensue. Hawaii demonstrates how peaceable, how successful, and how attractive in results this process can be. True, there are today conspicuous differences between the major races, but these are adaptations to conditions long different in their respective areas, and modern techniques give ready protection against these regional environmental difficulties.* As for generalized genetic advances, although the details call for much more factual study, it is evident that the gene leakage between areas would have allowed those migrant genes which presented universal advantages (even if they had had different times of entry and possibly somewhat different ratios of selection after that) to have become selected virtually everywhere.

Similarly, social origins must not be regarded as valid clues to

* Although the late 1960s saw a rejection of the melting-pot philosophy and a return to ethnic and racial identification, the miscegenation already in process throughout the world is likely to continue. Muller's view that miscegenation leads to new racial forms without biological impairment is accepted by most geneticists. In Hawaii and Latin America such wide-scale miscegenation has become an accepted pattern of marriage. In North America miscegenation is still subject to social discrimination.—Ed.

genetic level for classes with different status are all highly heterogeneous genetically. In the case of both ethnic and social or economic classes, however, the cultural differences are often so great as to give an utterly false appearance of genetic differences. At the same time, it must not be denied that the extent and importance of genetic differences between *individuals within any group* are often enormous, though, here too, any given person will have been shaped by environmental differences to a degree which would elude reliable assessment today.

It is generally conceded that the advances of science and technology already carry the physical potential of bringing dignity, affluence, health, enlightenment, and brotherhood within the reach of all. It is also conceded that, because of the dearth of really integrative and cooperative thinking and the inertia of old ways, these very advances are misused to cause the desperate crises of fast mounting population, massive depletion of resources, mass pollution, maldistribution, mass want that knows it need not exist, inflexible privilege, mass miseducation along outgrown lines, mass deception, frenzied fanaticism, mass coercion, the threat or actuality of mass slaughter and the destruction of civilization.

Thus, the changes in social conditions depicted in the earlier paragraphs constitute, all taken together, no more than a now-foreseeable larger cultural step forward which, as informed realists everywhere are aware, has become mandatory for the survival of civilization. It can bring no utopia—there will never be such a stasis, it is to be hoped!—but it will herald, in a sense, only a beginning of progress on a somewhat less insecure basis.

Man Undermines Himself

In considering these matters, we have not really strayed from our original subject. For, as we have seen, our genetics and our culture are inextricably interrelated. Only after such a general review could one informedly consider man's genetic future.

It might here be objected: "Why be concerned about our genetic future at all since, if civilization does continue, our scientific culture is likely to advance so fast as to much more than make up for genetic shortcomings—especially those affecting mental traits, in view of the enormous plasticity of man's mind? The means will be found," it is declared, "of suitably affecting gene action, immediate and remote, through methods of DNA and RNA repression and stimulation and other influences on the phenotype, or so-called euphenics, which should of course be considered as including ontogenetic influences, medicine, prosthetics, education, and applied sociology." In answer, I for one would thoroughly agree that efforts of these kinds ought to be actively pursued, and that some are bound to be highly rewarding both theoretically and practically. And certainly our minds do have immense hardly tapped reserves that could be made far more available just by the more suitable organization of both our early and our later experiences. The most immediately promising of these fields is of course educational reform.

Nevertheless, it would be utterly unrealistic to ignore the genetic side, in relation to either body or mind. If in this argument the antigeneticists who pose as supergeneticists were right, they would have to admit that their future methods could, by considerable expenditure at least, also convert apes, or, for that matter, any forms much less advanced than apes, into the equivalent of the most advanced people. Moreover, if genetic defects and shortcomings were to be allowed to accumulate to an unlimited extent among us, as seems to be happening now, the condition would eventually be reached in which each person likewise would present an immense, yet in his case distinctive, complex of problems of diagnosis and treatment.

But why then start with organisms? As I have pointed out before, the designing, once and for all, and the manufacturing of robots of choice from inorganic materials should prove much simpler. It is a possibility not to be scoffed at in this context, especially if people who have achieved such advanced techniques still insist on the *mystique* that the younger generation in each human family must be physical continuations of the older ones and must carry, if not their defects, at any rate a random sample of their oddities.

Just as natural mutations had to be stringently sifted by natural selection if a population were to advance or even not to deteriorate, so, in species divided into many small groups the mutational combinations in each had to be sifted, by a longer-range natural selection, in the interests of the species as a whole. And again, genera with only one species had, other things being equal, less chance of surviving than did multispecific ones, since any single species is so likely to prove, in the still longer run, to have been a natural error. This is shown by the fact that such a minute percent of species of the past have turned out to represent lines that persisted. In accord with this principle is the finding that the category with the highest percent of survivals has been that of phyla, and that successively narrower categories have had a correspondingly decreasing survival rate.

All this might have been expected, since some evolutionary trends are in directions that are later blocked by changes in other species or in the physical environment. In addition, as both Haldane and I long ago pointed out independently, some individuals which are technically fitter, in that they multiply faster than the rest of the population, do so at the expense of the rest and thus sap that population as a whole. Here, a division of the species into many small groups would tend to save it by weeding such groups out before the trait spread.

In the case of man, it has been intrinsically dangerous for him to have so long existed as just one species. He has been saved not only by his unparalleled advantages but also by having until recently been divided into thousands of tiny bands of at most a few score members each. In fact, as we have seen, this condition was especially favorable for the genetic enhancement of cooperative traits, including, I might add, those promoting group initiative or even—to use a harsher word—aggression. Until some two hundred generations ago the population pattern remained like this over by far the largest portion of the area inhabited by man. However, the agricultural revolution resulted in larger, denser, fewer groups, and the urban revolution greatly intensified this trend, thus practically preventing further genetic advances based on intergroup competition and even, in all probability, threatening the maintenance of those previously gained. This must be all the more true in the world of today and tomorrow.

At the same time, intragroup natural selection, working via families and via individuals, is also counteracted, as much as our improving techniques can, by saving everyone whom they can for survival and for reproduction. They have already become highly effective in this job. This means that mutations having a net detrimental effect on body or mind may now be accumulating almost as fast as they arise. We can escape the inference that such mutations far outweigh any advantageous ones only by believing that mutations are designed by Providence for a species' direct benefit, but in that case we run contrary to the clear experimental results.

It has, however, been suggested that even without the "survival of the fittest" that operated on man in past times there are features of modern conditions which cause enough differential reproduction, not consciously directed, to maintain or possibly advance the genetic constitution. But the data offered as evidence, which deal with mental ability (as gauged by IQ) or psychopathic deviation, are unsatisfactorily meager in just those parts of their range—the extremes—where they seem to indicate such an effect. Moreover, out-of-wedlock offspring are not mentioned, and they might have far more than offset any result of the kind supposed.

Although larger amounts of more detailed data on these matters should certainly be sought, as by the kinds of statistics concerning whole populations urged by Newcombe, whatever trends may be found are likely to become outdated rather soon. For techniques, social conditions, and attitudes throughout the world are changing so complexly and rapidly as to give such trends a seemingly erratic course. More trustworthy conclusions regarding long-term trends may therefore be reached on the basis of theoretical considerations rather than of extrapolated statistics.

As we have seen, these considerations show that modern culture by maximal saving of lives and fertility unaccompanied by a conscious planning which takes the genetic effects of this policy into account, must protect mutations detrimental to bodily vigor, intelligence, or social predispositions. Hence it must allow more accumulation of detrimentals in populations than would otherwise be the case. It appears wishful thinking to suppose that there is in our type of culture a built-in selective mechanism, not designed by us intentionally, which acts over a long period so as adequately to replace the earlier positive feedback whereby the genetic constitution was advanced.

Yet degeneration by passive accumulation of mutant genes is extremely gradual in its manifestation. This is the case even though approximately one individual in five can be reckoned to have received a detrimental mutant gene which arose in one of his parents and is added to the much larger number which they passed on to him from earlier generations. The reason for this creeping pace is that most mutant genes exert such minute effects, at least when the given gene has been received from only one parent. Slow also would probably be the effect of an automatic selection by modern culture in the direction of a lower level of physical or mental fitness. Hence these problems of creeping genetic deterioration are not acute in comparison with the fast-growing menaces presented by our cultural imbalances.

The much more important genetic problem arising out of modern cultural conditions lies in the need for further *advance* in the genetic level of those psychological endowments which have already attained a height so distinctive of man. These, we have seen, are cooperativeness and general intelligence, including the creativity which arises from initiative working through intelligence.

Why More Intrinsic Cooperativeness Is Urgent

Let us consider cooperativeness first. A stronger, broader cooperation is becoming imperative for adjusting to the relatively new conditions of life in large communities, and especially in the hoped-for world community of equal opportunities. Even in the scant two- to four-hundred generations since the ancestors of most people gave up living in tiny bands there may have been some significant passive accumulation of retrograde genetic changes that adversely affected one's brotherly feelings toward one's more distant associates. Under the dog-eat-dog mores prevalent, unofficially, within some of the larger, later communities there may have been an actual selection downward in such respects.

At any rate, the inadequacy for large communities of the level finally reached is indicated by the repeated and forceful entreaties on behalf of a broader brotherliness put forth by both past and present leaders of nearly every influential religion or ideology. We know of them from widely scattered regions, from soon after the time when diverse peoples there had been brought under a common rule. Yet, though they have fitted-in to some degree with most people's deep feelings, they have never in these millenia sunk-in thoroughly enough to be acted on without great inner resistance and outer friction.

So-called enlightened self-interest is no substitute. It can lead people in communities already having socially oriented practices to conform, outwardly at least, though it alone cannot initiate such community practices. But these same conformists, including those of high intelligence, may on feeling safe from exposure engage in unfair, cut-throat competition, covert fraud, or more extreme criminality.

More modern means of bringing up the young and of influencing the mind will doubtless be much more effective in the development of social consciousness and behavior and the repression of antagonism. Yet we are far from knowing to what extent this influence would be able to rival or exceed a deep and broad warmheartedness which was genetically built-in. That is, we cannot now estimate success in evoking from a man genuinely cordial reactions toward outsiders, utilization of his potentials, and affording him as much of a sense of fruition, as from one of those persons, at present so rare, whose genetic constitution has gone far in these respects. Meanwhile, the exigencies of recent culture call on us not to leave a stone unturned that could cause more of the population to be of this predisposition.

That differences in genetic constitution, not only in upbringing, between *individuals* of the same group are highly important in evoking social feelings and behavior, has long been objectively evident. This is now reported to be marked even among wolves. Such differences between species, of types which practically lack culture, are enormous, as are those between artificially selected breeds of dogs and of other domestic animals. Of course, any character which shows differences between species or breeds discloses similar differences: some large but with a range of sizes;

most of them small—between individuals of a given population. Moreover, Shields's studies of genetically identical twins reared apart, as compared with those reared together and with non-identical twins or other sibs, demonstrate that in man this principle holds for the other personality traits which he investigated.

But the avoidance of disaster should be far from man's only motivation in seeking a stronger, broader brotherly love. Many of us realize the truth behind the saying "Love is what makes the world go round." Since such feelings and behavior have already been built into our genetic constitutions and built-up in our cultures to a considerable, although not now sufficient, degree, we do appreciate and seek them, even for their own sakes. Built-in and built-up also has been the related gratification at knowing ourselves and our associates to be cospectators and coworkers forming a part of some much greater phenomenon than that of serving our individual self-interests. And, of course, both our basic feelings and our major cultural tendencies fit in with all trends which, like these, have long been of survival value in our line of ancestry. It is clear that we have in this way been brought to a consciousness that brotherly love will at the same time promote our survival, help to remove the aimlessness and sense of alienation so prevalent today, and afford man deeper inner fulfillment in working for his own vast community.

At the same time, we must maintain if not increase our comparatively high initiative or, in a sense, aggressiveness, but employ it hereafter in the form of independence of judgment and moral courage which, contrary to the ways of the so-called "organization man," braves social disapproval in order to start or support social and moral reforms. This initiative is also essential in our joint war to overcome the difficulties of nature. Moreover, it can certainly be exerted usefully in competition with one's fellows, so long as the competition is honest and constructive. It is evident, however, that its combination with sincere cooperation is indispensable in directing these activities toward the service of society rather than toward mere self-interest. When in combination with intelligence, initiative might better be termed creativity, but the combination of all three is required for the creativity that man needs.

On the Need for Increased Innate Intelligence

As for intelligence, consider how lost most people are today if they try to grapple realistically with our bewildering ideological, social, technical, or scientific problems. In all these areas more background, insight, and integrative ability are fast becoming required. Meanwhile, it is also becoming increasingly important that the politico-economic system be such as to seek from everyone his sincere, informed voice in the determination of policies affecting himself and his narrower group. These would of course include general policies too. Otherwise, fanatical individuals and cabals, avid for power, can too easily come to control and misuse mass methods of influencing the mind and of coercion, and by these means precipitate worldwide catastrophe. Yet a person's voice is worse than useless if it is not informed and understanding.

Here again better education, using the term in its broadest sense, can be of enormous help, and future biomedical and biochemical methods might also go a long way. But similar considerations apply here to those discussed in the case of cooperative feelings: the genetic constraints, though stretchable, are very real ones. Within any given social group of our own countrymen, the inter-individual differences in presently measurable features of intelligence are found to be based at least twice as much in genetic as in cultural differences and to give a range from idiocy to generalized genius. But suppose we might someday exert considerable *nongenetic* influence to raise the level. Should we now count on this to satisfy our need? Personally, if I had an opportunity to gain greater intelligence or understanding I would take advantage of any and *all* means of doing so, except, like Faust, had I to sell my soul. Moreover, our species as a whole for a very long time made the same choice, even though unconsciously. Hence our own dominance.

In fact, in consequence of the long-continued genetic selection in that direction, intelligence and probably cooperativeness are traits which would allow artificial selection of their positive extremes without (or with minimal) upsets in other respects. This conclusion is verified by the relatively high level of vigor and other valuable attributes which accompany these extremes and by the positive correlations among nearly all these traits.

But in spite of the continuous advance which natural selection has caused in our intelligence during the past two million years, at least as attested by one of its many factors—brain size—the rate of that advance, so amazingly rapid for mammalian evolutionary change, would be pitifully inadequate for us now if we could restore it. It is easy to reckon that, during the entire interval over which this rise in brain size was spread, the increase averaged not quite a tenth of one percent per thousand years. This was also the percentage increase during that part of the interval which lasted from the man who is first known to have used fire—the Peking man of some 400,000 years ago—until the man of today. This means an absolute increase that averaged not quite one gram per thousand years, from Peking man to our modern man whose brain weighs 1,400 or so grams.

There seems no reason that there need be any limit, except that set by our intelligence, to the advances made in science and technology and to the creative powers they would allow us to exert. Nor do we now see any necessary limit to intelligence, although great increases in human intelligence would doubtless require, at times, breakthroughs released by anatomical and biochemical innovations in the brain or its accessories. Such innovations (e.g., the *corpus callosum*) have taken place in past mammalian genetic evolution. In culture, there have been analogous ones (e.g., writing).

The present advances in our techniques and the consequent increases in the complexity of our social organization have, as we have seen, created a situation which demands much more understanding than people at large can muster. Hence they tend, nowadays, to become mere cogs in the mechanism, as well as to be too dependent upon it. It has sometimes been held, however, that this situation is a necessary evil, since there is today no place for many highly intelligent people: they find themselves doing work that is too routine for them. But the contrary is potentially the case, now that so much of the routine can be left to machines. For by these means an increasing number of people can be left free both to attain greater understanding and also to carry on higher work in general, including varied creative activities. That is, the intelligent can find more ways to utilize automation and other complex technical and social organizations so as to proceed to higher enterprises, which hold for them more fulfillment.

Perhaps the person who even today finds his work most gratifying, and who least feels the need of distractions such as hobbies, is the scientist, provided he is not hounded by too much competitive pressure to obtain and exhibit, with maximum speed, masses of novel-appearing but perhaps ill-considered data. Intrinsically, man is made in such a way that if he is intelligent science can be play for him. For, after all, it is really a superb expression, developed in cultural evolution, of primate curiosity and love of variety. In the future, if people in general can come to have a greater amount of creative intelligence they can partake of this fun in larger measure, both on their own, in groups, and, by learning about the work of others, vicariously.

Even when scientists seem to be on their own they are able to pluck their fruits only because they are privileged to be standing at the top of a vast pyramid of mental athletes of past and present. Would it not be better if just about everyone were so constituted that he could share in the joy of understanding the great collective conquests of his species, such as mathematical relations, relativity and its development, cosmogony as known at the time, biochemical evolution, exobiology, mind-body relations, intra- and intermind workings, social and industrial organizations, the latest artificial mechanisms, and so on, instead of having such understanding necessarily confined to a rare few, and compartmented among them at that?

Science and its technologies are unprecedentedly great enterprises of cooperation in intelligent creativity resulting from the cultural evolution of our genetically gifted species. They not only foster, when duly integrated, man's greater expansion and security but also give deeper fulfillment to man's own inner urges, both as built in him genetically and as further shaped culturally. They should therefore be, for us, at the same time the means of advancing our species and also their own excuse for being. Their development is soundest when the validity of either motivation by itself is recognized, as well as that of both working together. We would be false to ourselves if we did not try to promote throughout the population of future generations, by all possible means and without limit, the intelligence and the motivation which encourage the further advance of these activities. Therewith would also come the flowering of beauty: mainly in forms very unlike those of past or present but, utilizing the more advanced means and fitting in with the outlook, appropriate to the world which we will then be finding and creating, thereby enriching us both in our rest and our work-play.

Ways of Working Toward the Major Aims

The most basic way of working toward the major aims is to educate everyone not later than in high school in the main principles of biology, including especially genetic and cultural evolution and their lessons for ourselves. On the heels of this should be a sketch of world history, depicting the growing unity of man.

Even with all this background, most genetically less-fit individuals would not accept the judgment of their being so themselves and then voluntarily engage in less than the average amount of reproduction. Nor, vice versa, would the more fit choose to make the career sacrifices today made likely by their having larger families. True, this difficulty could be taken care of on paper, by taxes and subsidies, but in a democracy the enforcement of such seeming discrimination would hardly be accepted. Also rejected as discrimination would be known arrangements whereby persons of types prejudged less fit were automatically shunted (even though induced by the "carrot method") into ways of life otherwise attractive to them but affording less opportunity to reproduce, while those of types prejudged fitter had occupations that had been adjusted to provide greater opportunity to have children at little or no sacrifice, or with a subsidy.

However, with the educational background outlined, increasingly large numbers of couples who were suffering from sterility in the husband would be eager to avail themselves of means of having one or more children derived on the male side from someone they both held in deepest regard as a person physically by no means inferior while morally and mentally outstanding. There are perhaps ten thousand children a year produced in this country by artificial insemination with semen from donors chosen by the physician; but he does not select them according to such standards and he keeps their identity secret from everyone, including the couple. Well-endowed children would be far more desired if the couples were allowed to exercise the deciding voice in the choice of the genetic father after seeing the records concerning a wide range of possibilities, considering counsel concerning them, and judging which of them have shown more of the traits preferred by the couples themselves. Are not fertile couples nowadays expected to make their own choices of their partners in marriage, and are they not in that way allowed to choose also—even though with far less directness or likelihood of getting what they prefer than by the method here proposed—the kind of children whom they themselves want?

Openness of choice regarding donors would make it desirable that the semen had been stored, preferably for decades, until after the donors' decease. Thus the disclosure of the fact that a given person had been the donor could no longer handicap him nor present the possibility of personal entanglements between him and the recipient couple. Moreover, perspective could better be gained on the possible donors' phenotypically expressed merits and their genetic reliability in passing these along—information which would be invaluable in the making of choices.

Gradually, increasing numbers of nonsterile couples also would want to take advantage of so attractive an opportunity, for at least one child in their family. The previously mentioned resistance to self-condemnation which would interfere with voluntary eugenic action of an old-style kind would not operate here, since one would be comparing oneself with a personality whom one felt to be really great. The first participants would be those wanting a child without some particular defect of the husband's and idealistic realists who were far from subnormal. For the latter, clearly, quite open choices made voluntarily but after counseling and considering of the documentary evidence, would be essential. Then later, others would be proud to follow suit, letting it be known that they had done so.

There are additional reasons against secrecy. One is that adopted children usually find out that they have been adopted, as would "half-adopted" ones (to use Julian Huxley's term). The adopted child's attempt to discover his genetic derivation when (as is now usual) it is a closely guarded secret, commonly acts like a cancer in his life. On the other hand, in cases of the sort here described, knowledge of the facts would exert the opposite influence. Moreover, due appraisal of the data actually *requires* genetic recording of an open type. So does the making of genetic judgments about the future possibilities of an individual's germ cells, as well as the avoidance of incest, when the time comes for any given child to reproduce.

Of course the couples would be warned beforehand that genetic segregation and environmental influences allow the results of no human reproduction to be predicted, and that such selection as here depicted only *weights* the results in their favor. It would however be pointed out that outstandingly good performance has almost always required a combination of both favorable environment *and* favorable heredity; also that one-half of the child's non-sex-linked genes are those of the donor father. That the environment of these children also would tend to be favorable is indicated by follow-up studies on the families of those sterile couples who even today have resorted to artificial insemination, for their marriage and family life have turned out to be actually improved, on the average.

Nevertheless, there should be noted here the qualifying phenomenon of "regression toward the mean." Measurements of all sorts of traits in all sorts of organisms have long been known to show that the magnitude of any trait in an offspring tends to be nearer to that of the population's mean than just halfway between that of the more extreme parent and the less extreme one. Considering only cases in which the more extreme parent is on the "plus" or "positive" (useful or beneficial) side, and the less extreme one either "plus" or "minus," the genetic reasons are probably as follows: (1) The genes of the more plus parent (whom we'll say was the male) formed a uniquely favorable *combination,* in relation to the environment in which he was raised and maintained. So much was this the case that a random sample of half of them would, oftener than not, tend to operate somewhat less favorably in relation to their environment. Hence they tend to afford somewhat less than their "expected" half share of the effect in the offspring. (2) It is likely that extremely deviant genes (whether plus or minus) tend to have less dominance than genes closer to the normal in their effects.

As far as environmental effects alone are concerned, even when the genetic composition is identical in both the parents and all their offspring, as in a homozygous stock, similar considerations operate, since the more plus individual has usually had a combination of environmental factors operating in his behalf, such that their dilution by half or by any given proportion would be likely to result in a somewhat less than proportionate plus effect. Among human beings, both the above genetic and the environmental mechanisms mentioned would be operating at once. In addition, the highly favorable matching with one another of the more extremely plus parent's genetic and the environmental combinations would tend to be less than half attained in the offspring, regardless of the other parent's contributions to the genetic and the environmental circumstances.

Natural and artificial selection, however, have both operated successfully, though somewhat more slowly than otherwise, using the performance test (even when that of the individual alone is used). Selection based partly on the relatives' performance succeeds with a good deal less regression. Nevertheless, in terms of gene content, the "upgrading" does remain 50 percent, and very real progress is evident. In general, the more it is a matter of individual dominant genes of large effect, the less regression there can be, and regression is also reduced when the given gene tends to express itself with about as much relative strength when the accompanying group of genes and environment have a generally favorable or unfavorable phenotypic effect. Here the gene in question does tend to vary in its expression but in the main only *pari passu* with the genes in general. (Thus such a gene tends, at the *very least,* to be expressed with similar sign on "backgrounds" in general, as has been true of so many genes as to have maintained sexual recombination—a matter explained in other papers.)

At any rate, regression toward the mean may be regarded as traceable entirely to phenomena of the gene's *expression* varying with its environment. This may be its "exo-environment" or its environment of genes at other loci, or that determined by the kind of allele it has. In the genes the matter is purely one of pheno- or developmental genetics, not of heredity "proper." The time is exceedingly remote when the intricacies of all the specific interrelations here involved can be ascertained, even by the best computers. Meanwhile, we can *and must* improve greatly on nature's and the breeder's most successful attempts by using their performance criterion with the modern far more advanced techniques and the best pooled foresight that are now available to us.

Then, as the results, so favorable on the whole, of the relatively few first trials gradually become known, ever more couples will want to follow these pioneers' example; that is how new customs usually start. Previous taboos against the practice will dwindle. In their place, a new atmosphere of hope will emerge: hope both for the rewarding results likely to accrue to the couples themselves, and hope among them and others for mankind in general. Thus a genetic leaven will tend to diffuse through the population, and also a cultural, spiritual leaven. At last *human* resources, even on the genetic side, will begin to be enhanced, at an accelerating pace.

Admittedly, all this selection will be empirical, that is, based on performance rather than genetic analysis. After all, performance has been the criterion by which nature effected all our past evolution. Human discrimination, refined with the help of intelligent counselors, can however result in a far faster upgrading than nature has ever achieved in higher organisms, especially in regard to the two major characters that are being stressed. For considerable allowance can be made for the interfering effects of environment and of modifying genes, and heritability as indicated by close relatives can be taken into account. Knowledge of the actual genes concerned here is far from essential, however, and may be rather distant; each trait may well have many similarly acting major enhancers.

Whether or not, or how soon, we could supplant the chiefly empirical method with what I have termed "genetic surgery," we must not wait for this contingency to arrive, in view of our having the empirical method already at hand to some extent. For the application of the ultrasophisticated techniques to the kind of characters here being considered is likely to come much later than their more optimistic present supporters imply. Similarly, we should not wait for the success of so-called euphenics in this area.

Although success in upgrading the two major characters requires the high value of them to be recognized by the volunteering couples who engage in the project, they can and should have a great diversity of preference with regard to other characters. At the same time, they must be willing to take into consideration the counsel given them; one of the functions would be to see to it that there actually was enough diversity in their choices of other traits. In this way, moreover, such faddism as a flocking to choose as donors some then-popular group—such as certain types of entertainers, exhibitionists, sports champions, or demagogues—would be held in check.

Despite the differences in choice among couples, they would wish, and should be guided, to include some of the more special gifts of predilections which tend to support or channel the two major ones of cooperative disposition and general intelligence aimed at by all. Among these are: joy of life, strong feelings combined with good emotional self-control and balance, the humility to be corrected and self-corrected without rancor, empathy, thrill at beholding and at serving in a greater cause than one's self-interest, fortitude, patience, resilience, perceptivity, sensitivities, and gifts of musical or other artistic types, expressivity, curiosity, love of problem solving, and diverse special intellectual activities and drives. This list is very incomplete, the traits are complex, and many overlap and are interdependent. Some can be too extreme and should be held in temperate measure; some have bad names also (e.g., recklessness for fortitude). Many are greatly influenced by past environment. Physical traits also (e.g., longevity, late senility, vigor, good autonomic regulation, agility) should be given considerable place. No one has nearly all these mental and physical endowments, but that choice should be made which, while largely consistent with the counsel, best fits the couple's ideals. Initiative was omitted above because outstanding persons usually have much of it anyway, and those with too much "push" tend to become too strongly represented.

As these more special gifts become commoner in the population they can and should be more and more combined. This process will not ultimately reduce diversity. For the resulting population of more generally well-endowed individuals will of course branch out again diversely from the higher general level so attained. Thus it will gain still greater aptitudes of varied kinds in its different members.

Since really outstanding persons are relatively few, semen from the same donor would have to be used for correspondingly many people. Apprehension has been expressed that the genetic diversity of the population might become too much diminished in this way, but a little consideration shows that this supposition is incorrect. For one thing, the genetic spread of even a relatively small group of donors taken among the general population would be just as great as in the population in general in regard to other characters than the two major ones agreed upon, except for representing more of the really desirable "minor" characters above referred to, and even in regard to these and the two "major" characters, the phenotypic convergence would not necessarily imply a great deal of genetic similarity.

Moreover, our modern populations can be regarded as having arisen by a rapidly repeated doubling that occurred in a relatively small number of generations; during that time the newly added mutations were very few as compared with the ones that were carried along from the past, when the populations were a small fraction of ours today. In other words, our present genetic diversity is really that of a fairly small population, but the diversity is many times repeated in essentially the same way. Thus, the population could be subjected to generations of back-crossing, even to very few individuals in each generation, before it began to have its original *potential* diversity substantially reduced. But long before it was, the aims of the preliminary upgrading would have been achieved, and the improved population thus resulting would have devised better methods, of varied kinds.

In getting this project started, it is of the utmost importance that rigorous precautions be taken to insure that the persons in the group or groups taking part genuinely understand and favor the two major aims previously stressed. Persons who favor what *they* consider genetic improvement are of course all agreed on the major value of intelligence. However, they are far from agreed on the need for more cooperativeness, and even of those who believe they favor it a large number are gravely mistaken about its nature. That is one reason it has here been placed first, before intelligence. Many persons would consider as desirable cooperation today joint actions that would give preference to their own race, or nation, or class, or institution, or religious or provincial group, rather than to mankind as a whole. I do not mean by this to imply that mankind as a whole might never be served by taking sides in a dispute—far from it—but I do mean that a consistent policy of favoring your own side just because it is your own is contrary to the kind of cooperation needed in today's world. I have plenty of evidence that people would try to get into a project like the one here outlined and then channel it into this narrower concept in which it would be actually defeating its original aims.

Thus the group of prime-movers must be small and carefully chosen and guided by rules that maximally safeguard their future observance of this interpretation of social values. They will have to see to it that both the tasks of getting the material and of choosing the counselors are carried out in the same spirit. Although it would doubtless be helpful to have governmental blessing for the major aims and methods of such a project, actual governmental direction or detailed control at this stage of world affairs would present too great a risk of partisan influence and also subjection to standards of excellence which were too bureaucratic.

The group of prime-movers should of course have as participants not only persons specialized in genetics, in the physiology of reproduction in its theoretical and medical aspects, in psychology, and in social sciences but also representatives from other truly humanistic fields. In this connection, it is important to note that I have found not a few religious or ideological leaders, of diverse commitments, adopting a not unfavorable attitude toward this project when it was explained. Included were representatives of the Catholic, Methodist, Baptist, and Unitarian-Universalist denominations, of Judaism, official Humanism, and Free-thinking. Moreover, persons brought up to Buddhism and Shintoism have expressed approval. Representation from several of these groups, widely spread, would be important both in its own right and for its influence on the public and its leaders.

Choice of counselors should of course be made by the same standards as those used in the choice of the prime-movers but less rigorously applied. So should the selection of the outstanding donors. These results would automatically follow from a good choice of alert prime-movers.

In regard to the attitude of the above groups toward intelligence, they should keep in mind that eminence and creative intelligence are far from the same thing, though usually confused. Truly creative intelligence is likely to break barriers that were previously observed. Therefore these creative, highly intelligent persons all too often fail to be recognized as such by their contemporaries, although they have a relatively better chance of being so recognized by the following or a still later generation. That is another reason for storing most of the semen for decades. But it also means that precautions should be taken to have in the group of those selecting the semen to be stored some persons who can better recognize creativity and who may therefore be somewhat suspect themselves. Moreover, not only majority opinion should count in the original selection of the semen donors but each person participating in the choosing should be allowed the option of selecting some donors even without the approval of any of his colleagues. None of the semen of donors in this latter group should be used currently; all of it should be stored. On the other hand, a specified small amount of semen should be available for immediate use, that is, during the life of the donor, at least at the beginning of the project. For in this way some persons already eager to use the method can set an early example, and some preliminary results can be obtained as a background of information and statistics.

Of course it is scientists—evolution-minded scientists—who will form the core of the prime-moving group. It is important for some of them to get behind such a project as soon as possible, recognizing and stressing the major aims. For it must be started soon enough to show by its example the importance of its purposes as distinguished from the aims of some other groups that will surely form very soon and try to get priority. There are all too many of the so-called social Darwinist and racist types, and those espousing a master-and-slave society, who are interested in starting such groups, or may already have done so, even in our own country. In addition, projects of this kind will certainly spring up soon in some other countries, and not all will have the desirable breadth of outlook. Thus there is sure to be a "germinal race," parallel with our weapons race, and in this race at least we can and must set the highest possible standards. Here we would have nothing to lose, but mankind would have everything to gain by our winning.

Although no one should be excluded from the prime-moving group, or the supporting one, just because of his official politics or religion, nevertheless there are political and also religious positions, seldom adopted by an entire denomination or political party however, which are incompatible with the truly social outlook needed in the project. Here we should not hesitate to decide against admitting persons having such attitudes, even though this policy may sometimes lead to our being accused of bias. Incidentally, one of the best ways of proving that our own aims are not narrow or biased is to have included in the material stored semen from donors of varied races and social groups and to raise no objection to any couple of a different race or group using such material if they want to, but they should never be pressured in either direction on this point. Of course the data gathered by those selecting the donors will include information as to race, social class, etc., for these matters often have much bearing on the environments they have had to contend with, or benefitted from, and therefore on the contribution from genetic sources.

The taking of extra precautions to insure a sound, forward-looking social attitude on the part of the prime-movers and supporters of the project of germinal choice is made especially important by the present mores of our American society. Although it is far advanced in social outlook and practices as compared with its condition of only a half-century ago—as my personal recollections can vividly attest—it has not yet advanced far enough along this road to make "performance," as measured by mundane success in our present society, a reliable clue to the possession of the two major traits here stressed. The reason for this is not only that nonconformity is still frowned upon but also because of the persistence of strong over-competitive, unsympathetic attitudes in our commercial Western culture.

Although the chief seeds of Western progress do lie in its science, technology, education, and struggling democracy, its most conspicuous spirit is after all that of raucous, hypocritical, and often misleading salesmanship, aided by vulgar display, along with mass distraction, petty politics, and a growing militarism.

It is unfortunate that military strength should still be paramount. Yet without it the West, along with all the world, is practically certain to be submerged in the spread of the much more outright deception and outwardly unquestioning conformism so commonly imposed on large areas lying on the other side of the ideological fence. At the same time, it should be recognized that many of the minor regions connected with the West are guilty of similar falsification and repression. Moreover, on both sides of the fence the leaders of the major regions, no matter how mistaken about the means they should use, have ultimate aims which are constructive and feel similar menace. Barring war, there should be gradual, salutary convergence.

In view of this situation, still so confused and subject to strong and dangerous currents and counter-currents, unusual vigilance will be indispensable for keeping the aims of the genetic betterment group here proposed from becoming perverted. Yet delay in beginning the project as an active operation becomes all the more dangerous. For rival groups will in any case get busy very soon, whose procedures use the same techniques but have aims which are already oriented in clearly nonsocial ways, or are inadequately guarded against perversion. It will be impossible to repress such groups everywhere and only a better organization that would at the same time command great respect from the socially more knowledgeable elements will fill the need thus created.

Techniques, Research, and Practical Aims

As is now so well known, human spermatozoa can be kept deep-frozen at the temperature of liquid nitrogen (and lower), without deterioration during prolonged storage, even though the processes of freezing and thawing still incapacitate a minority of them for fertilization. The addition of glycerin, and probably still better, DMSO, considerably reduces this undesired effect. At several places in this country, and in at least one abroad, banks of frozen sperm are already being kept. However, there has been no attempt so far as is known to procure for any of these banks the semen of donors who (at least by the standards here discussed) are outstanding. Moreover, the material is being used in a dictatorial, secret way by the physician, just as with the artificial inseminations that employ nonfrozen sperm. The infants from the deep-frozen sperm have been comparable in their normality with those from unfrozen (or quite untreated) sperm.

Deep-frozen sperm are less sensitive to radiation damage than those at room temperature, although it would of course be preferable to have them stored where there is likely to be minimal exposure to radiation and other potentially damaging influences. By means of modern cryogenic methods, the cost of maintenance of large numbers of samples over many years is becoming surprisingly low per sample. For this reason it will be all the more necessary to have the arrangements for keeping them provide for their permanent identifiability, free from the chance of their becoming confused with one another.

Research is badly needed concerning ways of "stretching" the amount of use possible for a given sample, since suitable diluents, long known for domestic animals, have not yet been found for man. Nor have ways yet been found of reliably fertilizing a human egg by a sperm *in vitro,* since some kind of sperm "capacitation" is needed beforehand, which normally occurs in the Fallopian tubes. Another need is to find out how immature germ cells that could of course be kept deep-frozen like other tissues, can be caused to develop into normal spermatozoa *in vitro*. This would make possible the unlimited use of a given sample of immature germ cells. In addition, of course, it will be desirable to investigate these problems in the case of female germ cells, which Sherman, using mice, was able to deep-freeze without injury—but in rare cases only—and now Burks, using human eggs, has applied the technique. Ways of flushing them out of the tubes without injury or operation must also be sought more actively, and so must answers to such related problems as those of parthenogenesis and nuclear transfer. Again it must be borne in mind that the groups with aims less guarded from social misdirection will surely be conducting research and development of a comparable kind.

The fear is sometimes expressed that since outstanding merit often fails to be recognized until late (if ever) in life, the object of the project might be defeated by the mutations accumulated in the sperm of the elderly. However, a little calculation readily shows that, at most, such mutations would be very few in comparison with the numbers already supplied to the sperm by earlier generations. Hence the positive merit of the individual chosen would usually much more than compensate for this relatively small increment.

However, the best way of meeting this as well as other problems of germinal choice would be to have sperm storage much more widely practiced by the youthful population, and this procedure would eventually make for wider range of choice. Moreover, if increasing knowledge had in the meantime resulted in any substantial changes in our standards of excellence, the repositories of germinal material would then be extensive enough to meet this situation. Ideally, almost every young man should have some of his germinal material stored, if only for his own sake and that of his spouse.

This consideration would apply especially to persons likely to have the germinal material in their bodies exposed to more influence from mutagens than is usual. Self-interest could dictate prior storage for persons in any work involving the danger of overexposure to radiation. This would include astronauts, the crews of supersonic planes, workers in nuclear submarines, etc. Even men not so threatened would often be glad to pay for such a service, just as a means of allowing them to be vasectomized for the achievement of reliable birth-control without loss of the potentiality of later reproducing if and when they so desired, and many wives would also favor such a method for their husbands. Thus the stores could eventually be increased enormously. However, a large, well-financed organization (though perhaps eventually supported by the fees of those who used these services) would in that case be necessary.

With changing mores regarding germinal choice these services would eventually present much more possibility of choice to recipient couples, and with more choice the mores would change more rapidly, in a self-accelerating cycle. Thus the preliminary choosing of donors would become largely unnecessary. All the more necessary, however, would be the adoption of means of insuring that the counselors, and those engaged in choosing germinal material for their families, recognize and truly understand the major aims here stressed. These points are being raised now because some recent developments have already presented them.

Prospects

If to some people such discussions seem "far out," it should be remembered that they deal with measures closer to realization than those of applied gene knowledge of the traditional kind, and ever so much closer than those of "genetic surgery." In fact, the latter procedures would also present the same weighty problem of aims to decide if they were not to be confined to matters which, though important, were by comparison trivialities. As we have seen, the germinal choice by empirical methods that is so much closer to realization still has to clear one or more technical hurdles of importance before it can be in very wide use. But these should prove readily negotiable if subjected to some concerted action.

In the empirical germinal choice project it is the matter of values that looms as paramount at present. This is especially the case because so many people who would like to be associated with the project fail to realize its importance, or what the major values should be, and they are therefore striving to get the techniques going, willy-nilly.

It may be objected that we should not glean values simply from what evolution has found as an aid to genetic survival in our line of ancestry, but we must seek them out where our hearts and souls, aided by our reason, inspire us to find them. The answer to this is that our hearts and souls and reason have been, in every sense, the most important products of our own evolution. This implies that ultimately, for man, it is true that *right makes might.* But it is also crystal clear that for our modern world the values here emphasized need to be further enhanced by all means available, both cultural and genetic. After we have succeeded substantially in this great undertaking we will be in a better position to make further decisions about our more distant genetic and cultural problems. Enough for the present to be aware that, as far ahead as we can now see, there is no necessary limit to our cultural or genetic evolution, even though man confronts such a crisis in these fields now, and probably will for quite a while to come. In this crisis he can destroy his civilization if he does not take the speediest yet most carefully considered action possible.

We, as geneticists concerned with man, should see it as a part of our own responsibility not only to enlighten the public but also to promote, in the meantime, the collection, documentation, and storage of superior germinal material. This would be that of men who, according to considered judgment, best represent the major aims of enhanced cooperativeness, based on more heartfelt, broader brotherly love and more creative and generalized intelligence. Only in this way can we meet the obligation we all have to the multitudes who have made us what we are—to use the insights they have afforded us in behalf of our successors.

We must avoid getting sidetracked into acceptance of the delaying procedure so prevalent in both academic and political circles, which declares: "This needs more study!" Of course it does, but it is clear that there are certain things which can and must be done at this point; also that some of us are the ones to do them, in collaboration with suitable persons in other fields, whom we must find and encourage. Chief among these immediate tasks is to start the practice of accumulating germinal stores and records derived from persons who so far as we can see embody the major traits here stressed. The example thus set will be the main feature of this starting effort.

By this stage of our discussion it should have become clear that there are certain dangers to be specifically guarded against in the choice of participants in the project and also in the selection of germinal material. One of the types most to be avoided as collaborators or as donors is that of the egotistical, paranoid personality who would certainly try to push or squirm his way in. No prime-mover, supporter, or donor should be a person who seeks the storage of his own germinal material. I know definitely of a physician who in practicing artificial insemination often used his own semen in the treatments. He remains safe under the cover of secrecy that dominates this field among today's physicians, and he had also saved money in this way. It happens that this physician was in fact intelligent, but his procedure illustrates that intelligence is not enough. A related type to be avoided as a prime-mover or supporter is the person who accepts as valid the over-individualistic, asocial viewpoint that is still so prevalent. The violation of these principles by any group engaged in an alleged human betterment program should, *if verified,* be exposed.

On the whole, physicians, especially those concerned with reproduction and the urinogenital organs, who would be willing to give up dictatorship and secrecy as principles to be adhered to in inseminations carried out under the germinal choice project, would be important to have as participants in the plan. But they are extremely hard to find. Meanwhile the work of getting the project going must be undertaken as soon as possible, and both medical and legal aid will eventually be forthcoming. In this connection, however, it should be borne in mind that general medical approval and adequate legalization seldom are accorded until after the practice itself has been initiated and has won approval in other highly regarded but more progressive circles. A good example here is that of contraception.

Thus we should not let ourselves be discouraged by the temporary difficulties. We should not only bear in mind the urgent need for success, we should also recall that, after all, man has gone from height to height, and that he is now in a position, if only he *will,* to transcend himself intentionally and thereby proceed to elevations yet unimagined. Unintentionally, he no longer can do so. It is up to us to do our bit in this purposive process and to use what we know constructively, rather than remain in that ivory tower which has the writing on its wall. Our reward will be that of helping man to gain the highest freedom possible—the finding of endless worlds both outside and inside himself, and the privilege of engaging in endless creation.

Summary

Our line of ancestry, after splitting off as the primate order from other mammals during the era of reptiles, was built for active tree life of a kind which gave an advantage in genetic survival to the little bands whose individuals felt more empathy for one another and better understood one another. Thus natural selection here favored genetic constitutions which resulted in more "social intelligence" and cooperation. The primate structure and mode of life also caused natural selection, apparently somewhat later, for more curiosity concerning objects, love of variety, and versatility, and thus for increased general intelligence. These aptitudes the apes carried still further, with their arm-swinging progression giving them a semierect posture and a wider view and also greater mobility.

In this way a trend toward a fully erect life on the ground was facilitated. It was pursued in a line of apes that split off from the others, some twenty million years ago, in a more definitely human direction. On the ground there was even more advantage in manual ability, improvisation, and cooperation—as in predation and in defense from predators. Thus these selective tendencies were accentuated. They gradually built up the intelligence and the collaboration between associates to such an extent as slowly to allow the evolution of culture, including speech. The increasing culture, in turn, provided conditions for further genetic selection among the small bands, thus stimulating the ability to better utilize and improve the culture. The major attributes thus favored were, again, cooperativeness and intelligence, aided by initiative, i.e., creative intelligence. This reciprocal positive feedback between cultural and genetic evolution must have lasted for some two million years, until about the start of the agricultural age.

At that time culture became so successful as to allow the groups of people to increase so in size that intergroup competition became virtually inoperative. Recently, success in saving nearly everyone for survival, and probably the great majority for full reproduction, must have gravely undermined nearly all genetic progress in the bases of our physical, mental, and moral natures.

By far the worst aspect of our present genetic crisis lies in the urgent need for further advances in the development of our mental and moral natures. First in importance should be the need for a greater capacity on the part of people in general to extend genuine warmheartedness, that is, brotherly love, to all sectors of humanity, in a *world-wide community*. Second would come the need to have our general creative intelligence deepened and broadened. By the latter means, the complex problems raised by our modern sciences and technologies, and our resulting social organizations, could be far better resolved by us. They could also be better understood by mankind in general, so that people could live in awareness of the modern world which, after all, their forebears and contemporaries had discovered and created. Thereby their alienation and feeling of frustrated enslavement to a vast machine would be relieved. They could rise to savor the deep joys of sensing the great universe outside of and within them. They could eagerly join in the creation of ever better worlds, and rise over their machines to higher, less routine activities.

Genetic progress of these kinds will no longer occur automatically. We must use the advances of our modern culture to bring them about intentionally. Moreover, this aim should be pursued by every possible means, and we should also progress on a more purely phenotypic (nongenetic) level toward the same goal. These two methods of advance should not be regarded as antagonistic but as complementary. Moreover, in genetic progress we cannot wait until we have mastered some refined "genetic surgery," nor even until we have better knowledge of the individual genes concerned in the attributes we are aiming for—although of course that skill and knowledge should meanwhile be sought actively. We should at least make a start by using the empirical method—being guided by the test of performance which has worked in natural selection.

For this purpose a beginning should now be made in germinal choice. This means, for the present, of affording to couples the opportunity to have one or more children derived on the father's side from germinal material of their choosing, out of well-documented stores which had usually been kept for decades after the decease of those who donated them. The transactions must be recorded, not secret, and not dictated. Physical techniques are already available for conducting such an enterprise on a small scale. Areas of research to extend and improve these techniques, and the project as a whole, are outlined in this article.

It is important that means be employed, including counseling, to maximize the chance that the material chosen represents a genuine advance in the major genetic attributes we have stressed, while salutary diversity is maintained in other respects. Special precautions and vigilance are required to minimize subversion of the major aims, because our modern cultures are not yet sufficiently advanced to insure this. For this reason there is imminent danger of other groups (such as racists and espousers of a master-and-slave society), lacking awareness of social fallacies, soon making headway. Thus it becomes all the more necessary for practical work to be started at once by a more responsible, but nongovernment body, as a salutary example. What is called for is action by a socially oriented group of geneticists, backed by aides from the fields of physiology, psychology, and medicine and by diverse humanistically minded persons of competence, with the aim of beginning to establish and maintain germinal repositories intended mainly for use decades later.

Amidst the justified present concern about "natural resources" in general, our *human resources*—rooted in man's most basic nature—should by no means be neglected. It is hoped that in his present critical period of rapid cultural transition man will thereby be enabled better to evision and advance to ever greater realms of knowledge, creation, and spirit.

Chronology of H. J. Muller's Career

1890	Born in New York City, 21 December.
1904/1907	Attended Morris High School, Bronx, New York
1907/10	Undergraduate, Columbia University; influenced by E. B. Wilson for chromosome theory.
1910/11	M.A., Physiology Department, Columbia University.
1911/12	Assistant, Cornell Medical School.
1912/15	Ph.D. research with T. H. Morgan on crossing-over (coincidence and interference); independent work on gene and character relations; correlation of chromosome number and size with crossover maps and linkage groups.
1916/18	Recruited by Julian Huxley to Rice Institute; analysis of inconstant, variable traits.
1918/20	Instructor, Columbia University. Theoretical papers on mutation and the individual gene; reproduction of variations as key to definition of gene.
1921	Assistant Professor, University of Texas; analysis of complex traits with chief genes and modifiers.
1922	Develops ClB method for detecting sex linked lethals.
1923	Marries Jessie Marie Jacobs (divorced 1935).
1924	Son, David Eugene, born.

1925	Studies of identical twins raised apart.
1926	Gene as the basis of life throughout evolution.
1927	Artificial induction of mutations with X rays; AAAS Research Prize.
1928/32	Laws of radiation genetics; analysis of chromosome breakage as basis for chromosome rearrangements.
1932	Attack on American eugenics movement; disillusionment with contemporary American society; Guggenheim Fellowship to work in Berlin with Timoféeff-Ressovsky; analysis of dosage compensation.
1933/37	Senior geneticist, Leningrad and Moscow. Analysis of gene boundaries, gene size, viable deficiencies. Cytogenetic studies of Bar gene as duplication; gene evolution by tandem duplications. Polemic controversy with Lysenko.
1935	Publication of *Out of the Night.*
1937	Volunteer in Spanish Civil War, Canadian Blood unit; research on transfusion from cadavers.
1938/40	Lecturer, Institute for Animal Genetics, Edinburgh, Scotland. Studies of ultraviolet mutation; relation of chromosome changes to evolution; discovery of mutations induced with low rates of ionizing radiation.
1939	Marries Thea Kantorowicz.
1940/45	Visiting Professor, Amherst College. Studies of spontaneous mutation and difference in rates at various stages of meiosis; interspecific crosses and gene evolution in *Drosophila;* dosage compensation as evidence of neo-Darwinian mechanisms of evolution.
1945	Professor, Indiana University; Pilgrim trust lecture on the gene.
1944	Daughter, Helen Juliette, born.

1946	Nobel Prize in Medicine and Physiology.
1946/50	Lysenko controversy; concern over radiation hazards; concept of genetic load in man.
1950/57	Critic of medical, industrial, and military indifference to radiation damage; establishment of maximum permissible doses to human population. Banned by AEC as U.S. delegate to first *Atoms for Peace Conference,* Geneva, 1955.
1958/67	Advocate of positive eugenics; proposal for sperm banks; principle of germinal choice (voluntary basis for eugenic use of frozen sperm; donors selected for outstanding cooperativeness, intelligence and vigor to counteract man's load of mutations).
1967	Died 5 April, Indianapolis, Indiana.

Honors and Awards

Cleveland Research Prize, AAAS, 1927; Hon. D.Sc. University of Edinburgh, 1940; Nobel Prize in Physiology and Medicine, 1946; Hon. D.Sc., Columbia University, 1949; Distinguished Service Professor, Indiana University, 1953; Kimber Genetics Award, National Academy of Sciences, 1955; Rudolph Virchow Society of New York, Medal, 1956; Darwin Medal of Linnean Society, 1958; Hon. D.Sc., University of Chicago, 1959; Hon. M.D., Jefferson Medical College, 1963; Humanist of the Year, 1963; Hon. D.Sc., Swarthmore College, 1964; City of Hope National Research Citation, 1964.

Selected References

A. Books

1. Morgan, Thomas H.; Sturtevant, Alfred H.; Muller, Hermann J.; Bridges, Calvin B. *The Mechanism of Mendelian Heredity*. New York: Holt and Co., 1915.

2. H. J. Muller. *Out of the Night: A Biologist's View of the Future*. New York: Vanguard Press, 1935; London: V. Gallancz, 1936.

3. Muller, H. J.; Little, Clarence C.; Snyder, Lawrence H. *Genetics, Medicine, and Man*. Ithaca, New York: Cornell University Press, 1947.

4. Muller, H. J. *Studies in Genetics: The Selected Papers of H. J. Muller*. Bloomington, Indiana: Indiana University Press, 1962.

B. Obituaries and Evaluations

1. Auerbach, C. "Obituary Note, H. J. Muller." *Mutation Research* 5(1968): 201–7.

2. Carlson, E. A. ed. "H. J. Muller Memorial Issue." *Indiana University Review* 11, no. 1 (Fall 1968): 1–48.

3. Carlson, E. A. "H. J. Muller." *Genetics* 70 (1972): 1–30.

4. Pontecorvo, G. "Hermann Joseph Muller 1890–1967." *Biographical Memoirs of Fellows of the Royal Society* 14 (1968):

349–89. (Contains a complete bibliography of Muller's published works.)

5. Sonneborn, T. M. "H. J. Muller—Crusader for Human Betterment." *Science* 162 (1968): 772–76.

C. Source Materials at the Lilly Library, Indiana University, Bloomington

1. Carlson, E. A. "Indiana University: The Muller Archives." *The Mendel Newsletter*. Library of the American Philosophical Society, no. 4 (November 1969): 1–2.

Acknowledgments

Permission to reprint the essays in this volume is gratefully acknowledged. The following list gives the original facts of publication for all essays here republished.

"Possible Advances of the Next Hundred Years: A Biologist's View," was prepared for "The Next Hundred Years," a symposium held by the Seagram Company, 22 November 1957, at the Waldorf Astoria Hotel, New York City, on the occasion of its hundredth anniversary. Because of time limitations, the first paragraph was abridged and most of the material of the second to seventh paragraphs, inclusive, was omitted.

"Science Fiction as an Escape" is from *The Humanist* no. 6 (1957): 333–46.

"Life Forms to be Expected Elsewhere Than on Earth" is based on an address delivered by Muller at the National Association of Biology Teachers, 28 December 1959, Chicago, in connection with the annual meeting of the American Association for the Advancement of Science; it was also given in modified form at Taylor University 14 March 1961, Upland, Ind. It is here reprinted from *The American Biology Teacher* 23, no. 6 (October, 1961): 331–46.

"The Meaning of Freedom" is reprinted from *Bulletin of the Atomic Scientists* 16, no. 8 (October 1960): 311–16, copyright 1960 by The Educational Foundation for Nuclear Science Inc.

"The Radiation Danger" is reprinted from *The Colorado Quarterly* 6, no. 3 (Winter 1958): 229–54.

"In Search of Peace" is from *The Humanist* no. 2 (1959): 69–70. Copyright 1959 American Humanist Assn., Yellow Springs, Ohio.

"Human Values in Relation to Evolution" is reprinted from *Science* 127, no. 3299: 625–29.

"Social Biology and Population Improvement" is reprinted from Nature 144 (16 September 1939): 521.

"What Genetic Course Will Man Steer" was printed in *Proceedings of the Third International Congress of Genetics, 5–10 September 1966, Chicago, Ill.*, edited by James F. Crow and James V. Neel (Baltimore, Md.: The Johns Hopkins Press, 1967), pp. 521–43. The writer acknowledged with thanks support in the preparation of this paper afforded by grant number GB 4764 from the National Science Foundation to the Indiana University Foundation in behalf of his work.

Index